3rd International Workshop on Materials Analysis and Processing in Magnetic Fields 2008

(MAP3)

AA002628

Journal of Physics: Conference Series Volume 156

Tokyo, Japan
14-16 May 2008

Editors:

Yoshio Sakka
Shigeru Horii
Noriyuki Hirota
Tsutomu Ando

ISBN: 978-1-61738-377-9

Printed from e-media with permission by:

Curran Associates, Inc.
57 Morehouse Lane
Red Hook, NY 12571
www.proceedings.com

Some format issues inherent in the e-media version may also appear in this print version.

Copyright© (2009) by the Institute of Physics
All rights reserved.

Printed by Curran Associates, Inc. (2010)

For permission requests, please contact the Institute of Physics
at the address below.

Institute of Physics
Dirac House, Temple Back
Bristol BS1 6BE UK

Phone: 44 1 17 929 7481
Fax: 44 1 17 920 0979

techtracking@iop.org

TABLE OF CONTENTS

Magnetic Alignment of Poly(ethylene terephthalate) in Molten State 1
M Yamato, T Kimura, K Takahashi, K Watanabe

Three-dimensional Orientation of Cellulose Crystals Under Dynamic Elliptic Magnetic Field 6
F Kimura, T Kimura

Continuous Measurement of XRD Peak Intensity of Bismuth Under Magnetic Field 9
T Kohama, K Iwai

Different Properties Exhibited on the Two Typical Crystal Faces of Hydroxyapatite in a Simulated Body Environment 16
T Hagio, T Tanase, J Akiyama, K Iwai, S Asai

Alignment of Carbon Nanotubes by Magnetic Fields and Aqueous Dispersion 23
B K Jang, Y Sakka, S K Woo

Electro-migration Effect of Titanium Structure Under High Magnetic Field 29
Y Sakaida, K Iwai, S Asai

Rotational Diffusion Model of 2-dimensional Magnetic Alignment 34
M Yamaguchi, S Ozawa, K Ogawa, I Yamamoto

Crystal Alignment of Primary Phase During Solidification 40
K Iwai, M Usui, N Toyoda, S Asai

Texturing Hard Nd-Fe-B Powdered Ribbons in High Magnetic Field 45
S Rivoirard, V M T S Barthem, T Garcin, E Beaugnon, P E V de Miranda, D Givord

In situ Characterization of Phase Transformations in a Magnetic Field in Fe-Ni Alloys 52
T Garcin, S Rivoirard, E Beaugnon

Selective Formation of an Ordered Variant in $Fe_{45}Pd_{55}$ Alloy Under Magnetic Field 57
S Farjami, T Fukuda, T Kakeshita

Effect of Magnetic Field on Solid-solid Phase Transformations in Iron-based Ferromagnetic Alloys 61
T Kakeshita, T Fukuda

Effect of Magnetic Field on TTT Diagram of Successive $\gamma \rightarrow \varepsilon \rightarrow \acute{\alpha}$ Martensitic Transformation in SUS304L Stainless Steel 70
J H Lee, T Fukuda, T Kakeshita

Biological Effects of Strong Static Magnetic Fields on Insulin-secreting Cells 74
T Sakurai, J Miyakoshi

Effects of Static Magnetic Field on Mutagenesis in in vitro{i} 80
M Ikehata, S Yoshie, N Hirota, T Hayakawa

Yeast Cells Proliferation on Various Strong Static Magnetic Fields and Temperatures 84
E S Otabe, S Kuroki, J Nikawa, Y Matsumoto, T Ooba, K Kiso, H Hayashi

Effects of Gradient Magnetic Force and Diamagnetic Torque on Formation of Osteoclast-like Giant Cell 92
M Iwasaka, M Ikehata, N Hirota

X-Ray Powder Diffraction Camera for High-field Experiments 99
K Koyama, Y Mitsui, K Takahashi, K Watanabe

Development of an Extraction Type Magnetometer Under Low Temperature and High Magnetic Fields Over 20 T by the Hybrid Magnet 104
K Koyama, R Sakakura, K Watanabe

Precise Measurements of Diamagnetic Susceptibility of Benzophenone and Paraffin by Using a Magnetic Levitation Technique 110
K Takahashi, I Mogi, S Awaji, K Watanabe

Diamagnetic Susceptibility of Single Micro-particles Detected by Free Translational Motions in Field Gradient 115
K Hisayoshi, S Kano, C Uyeda

Ab Initio Quantum Chemical Calculation as a Tool of Evaluating Diamagnetic Susceptibility of Magnetically Levitating Substances 120
Y Fujiwara, Y Tanimoto

Attempt to Detect Diamagnetic Anisotropy of Oxides with Isotropic Crystal Structure by Measuring Its Rotational Oscillation in Strong Magnetic Field 128
S Kano, K Hisayoshi, T Kashiwagi, T Kida, M Hagiwara, C Uyeda

Attempt to Improve Sensitivity in Measuring Diamagnetic Anisotropy by Increasing Duration of Rotational Oscillation in Microgravity 133
C Uyeda, S Kano, K Hisayoshi, T Kashiwagi, T Kida, M Hagiwara

Optical Extinction Change of Magnetic Nanoparticle Suspension Under Pulsed Magnetic Field ... 137
M Suwa, H Watarai

Effects of Magnetic Processing on Electrochemical and Photoelectrochemical Properties of Electrodes Modified with Nanoclusters of a Phenothiazine-C_{60} Linked Compound ... 145
H Yonemura, Y Wakita, S Moribe, S Yamada, Y Fujiwara, Y Tanimoto

Chiral Electrode Behavior of Magneto-electrodeposited Cu-Cu_2O Films ... 155
I Mogi, K Watanabe

Magnetic Field Increases the Surface Tension of Water ... 161
Y Fujimura, M Iino

Effects of Gradient Magnetic Fields on CO_2 Sublimation in Dry Ice ... 166
M Iwasaka

Influence of Strong Magnetic Field on the Sedimentation of Red Blood Cells ... 171
Y Tanimoto, Y Kakuda

Study on Magnetic Separation System Using High Tc Superconducting Bulk Magnets for Water Purification Technique ... 175
T Oka, H Kanayama, K Tanaka, S Fukui, J Ogawa, T Sato, M Yamaguchi, M Ooizumi, K Yokoyama, K Noto

Study on Magnetic Separation of Nanosized Ferromagnetic Particles ... 180
R Nakao, Y Matuo, F Mishima, T Taguchi, S Nishijima

Removal and Recycle of Phosphate from Treated Water of Sewage Plants with Zirconium Ferrite Adsorbent by High Gradient Magnetic Separation ... 185
D Ito, K Nishimura, O Miura

Magnetic Separation of Antibiotics by Electrochemical Magnetic Seeding ... 188
I Ihara, K Toyoda, N Beneragama, K Umetsu

Author Index

PREFACE

3rd International Workshop on Materials Analysis and Processing in Materials Fields

Yoshio Sakka *et al* 2009 *J. Phys: Conf. Ser.*.

Yoshio Sakka, Noriyuki Hirota, Shigeru Horii and Tsutomu Ando

3rd International Workshop on Materials Analysis and Processing in Materials Fields (MAP3) was held on 14-16 May 2008 at the University of Tokyo, Japan. The first was held in March 2004 at the National High Magnetic Field Laboratory in Tallahassee, USA. Two years later the second took place in Grenoble, France. MAP3 was held as *The University of Tokyo International Symposium*, and jointly with *MANA Workshop on Materials Processing by External Stimulation*, and *JSPS CORE Program of Construction of the World Center on Electromagnetic Processing of Materials*. At the end of MAP3 it was decided that the next MAP4 will be held in Atlanta, USA in 2010.

Processing in magnetic fields is a rapidly expanding research area with a wide range of promising applications in materials science. MAP3 focus on the magnetic fields interactions involved in the study and processing of materials in all disciplines ranging from physics to chemistry and biology:

- Magnetic field effects on chemical, physical, and biological phenomena
- Magnetic field effects on electrochemical phenomena
- Magnetic field effects on thermodynamic phenomena
- Magnetic field effects on hydrodynamic phenomena
- Magnetic field effects on crystal growth
- Magnetic processing of materials
- Diamagnetic levitation
- Magneto-Archimedes effect
- Spin chemistry
- Application of magnetic fields to analytical chemistry
- Magnetic orientation
- Control of structure by magnetic fields
- Magnetic separation and purification
- Magnetic field-induced phase transitions
- Materials properties in high magnetic fields
- Development of NMR and MRI
- Medical application of magnetic fields
- Novel magnetic phenomena
- Physical property measurement by Magnetic fields
- High magnetic field generation

MAP3 had 84 presentations including 16 invited talks. This volume of Journal of Physics: Conference Series contains the proceeding of MAP3 with 34 papers that provide a scientific record of the topics covered by the conference with the special topics (13 papers) of the well indexed journal "Science and Technology of Advanced Materials". All articles have been refereed by expert in the field. Both of these journals are fully accessible electronically and can be cited and referenced in the usual way. It is our hope that the reader will enjoy and profit from the MAP3 Proceeding.

Chair

Hitoshi Wada (Kashiwa, Japan)

Co-Chair

Eric Beaugon (Grenoble, France)

Hans J. Schneider-Muntau (Tallahassee, USA)

Advisory Board

Shigeo Asai (Nagoya, Japan)

Koichi Kitazawa (Tokyo, Japan)

Mitsuhiro Motokawa (Sendai, Japan)

Shoogo Ueno (Fukuoka, Japan)

Robert Tournier (Grenoble, France)

Justin Schwartz (Tallahassee, USA)

J. C. Maan (Nijmegen, Netherland)

Scientific Committee

Yoshifumi Tanimoto (Hiroshima, Japan)

Masuhiro Yamaguchi (Yokohama, Japan)

Tsunehisa Kimura (Kyoto, Japan)

Yoshio Sakka (Tsukuba Japan)

Ryoichi Aogaki (Tokyo, Japan)

Jyunji Miyakoshi (Hirosaki, Japan)

Kazuo Watanabe (Sendai, Japan)

James M. Valles Jr. (Providence, USA)

Joon Pyo Park (Pohang, Korea)

Qiang Wang (Shenyang, China)

Nicole Pamme (Hull, UK)

Sophie Rivoirard (Grenoble, France)

P. C. M. Christianen (Nijmegen, Netherland)

Local Organizing Committee

Isao Yamamoto

Masafumi Yamato

Shigeru Horii

Norihito Sogoshi
Masateru Ikehata
Noriyuki Hirota
Tsutomu Ando

Proceeding Editorial Bord
Yoshio Sakka
Noriyuki Hirota
Shigeru Horii
Tsutomu Ando

This is a co-publication with *Science and Technology of Advanced Materials*. The bulk of the papers, after peer review, are published in *Journal of Physics: Conference Series*. However a selection of papers, are published separately in a special issue of *Science and Technology of Advanced Materials*.

All papers published in this volume of *Journal of Physics: Conference Series* have been peer reviewed through processes administered by the proceedings Editors. Reviews were conducted by expert referees to the professional and scientific standards expected of a proceedings journal published by IOP Publishing.

Magnetic Alignment of Poly(ethylene terephthalate) in Molten State

M Yamato[1], T Kimura[2], K Takahashi[3] and K Watanabe[3]

[1]Tokyo Metropolitan University, 1-1 Minami-ohsawa, Hachioji, Tokyo 192-0397, Japan
[2]Kyoto University, Oiwake-cho, Kitasirakawa, Sakyo-ku, Kyoto 606-8502, Japan
[3]Tohoku University, 2-1-1 Katahira, Aoba-ku, Sendai, Miyagi 980-8577, Japan

E-mail: yamato-masafumi@tmu.ac.jp

Abstract. The magnetic alignment of poly(ethylene terephthalate) in molten state was studied by *in-situ* birefringence measurement in a magnetic field. The melting point of the sample determined by DSC was 253℃. We observed the increase of birefringence in a molten state at temperatures between 254℃and 260℃ in the magnetic filed (10T). The dependence of the apparent rate of orientation on the annealing time was attributed to the increase in the viscosity due to the structure formation under the lower supercooling condition. The time at which the transmitting light became a maximum increased with the increase in the melting temperature. These results suggested that the residual structure of crystal, which can be aligned magnetically, existed in the molten state and the volume fraction of the residual structure was reduced with increase in the melting temperature.

1. Introduction

The physical properties of polymeric materials depend greatly on not only their chemical structure, but also their higher order structures. In particular, various physical properties depend on both the crystallinity and the degree of orientation in the case of crystalline polymers. It is possible to control the crystallinity by controlling the temperature conditions during the manufacturing molding process. However, controlling the orientation is not easy. The use of an external magnetic field is a technique that can be used to control the orientation of these materials [1]. Magnetic alignment can achieve unique orientations (such as graded orientations [2]) that cannot be realized by conventional methods.

In a series of studies [3-8], we have discovered that many crystalline polymers can align under a magnetic field. However, we have not reached a fully clear understanding about the factors that determine the magnetic alignment of crystalline polymers. One of the proposed mechanisms is the alignment of some anisotropic structure that might exist in the molten state [5, 6]. However, the magnetic alignment in the melt of polymers has not been observed directly. Moreover, it is reported that the magnetic alignment does not occur in the molten state of isotactic polystyrene [9]. In this study, we report the magnetic alignment of poly(ethylene terephthalate) (PET) in the melt by using an *in-situ* birefringence measurement in the magnetic field of 10T.

2. Experimental

2.1. Sample preparation
Pellet of poly(ethylene terephthalate) purchased from Scientific polymer products, inc. was used. Amorphous films of 150 μm thickness were prepared by cooling a melted sample in iced water rapidly after hot pressing. The sample (5 mm diameter) cut from the film were used in this study.

2.2. DSC measurement
DSC measurements were carried out using a SEIKO DSC200 equipped with a thermal analysis system SSC5200H under a dry nitrogen atmosphere. The melting point of the sample was measured at a scanning rate of 10 K/min.

2.3. Birefringence measurement
In-situ birefringence measurement was carried out with a home-built apparatus. The block diagram of the apparatus was shown in figure 1. A polarizer and an analyzer were set under the cross-Nicole condition, each making an angle of 45 ° with respect to the vertical magnetic field. The sample were set to the birefringence apparatus and allowed to melt inside and outside of the magnetic field. The change of the transmitting light intensity was measured under the cross-Nicole condition, which corresponds to the change of the birefringence.

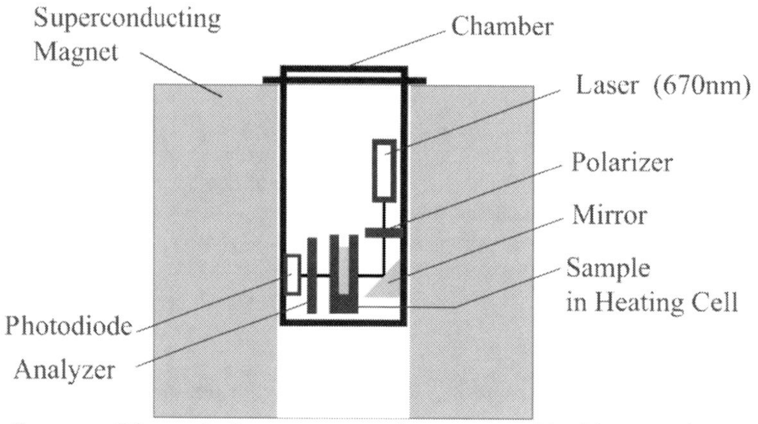

Figure 1. Block diagram of the optical apparatus used to measure birefringence in a magnetic field.

3. Results and discussions
The temporal change of the transmitting light intensity was measured both inside and outside of the applied magnetic field at an isothermal annealing temperature that is slightly higher than melting point of 253 °C determined by DSC measurement. Under the cross-Nicole condition, the transmitting light intensity is proportional to $\sin^2(\pi d\Delta n/\lambda)$, where d is the film thickness, Δn is the birefringence, and λ is the wavelength (670 nm) of the impinging light. The quantity $d\Delta n$ is referred to as retardation, denoted by *Re*. If the increase in Δn due to the magnetic alignment of the polymer chain is large, the intensity would even oscillate.

Figure 2 shows the change of the transmitting light intensity under the cross-Nicole condition during the isothermal annealing at 255 °C. No change in the transmitting light intensity was observed when the magnetic field was not applied, indicating that there was not a change in the birefringence. On the other hand, the transmitting light intensity oscillated as a function of the annealing time in the case of the measurement at 10 T, suggesting the increase in birefringence. This result indicates that some anisotropic structure appeared in polymer melt. We succeeded in detecting the magnetic alignment in the polymer melt. The structure aligned in the magnetic field is considered as some residual structure of crystal that existed before melting [5, 6]. In addition, the cycle of the oscillation increases as time progresses. This result suggests that the rate of orientation decreases with time. Possible reasons for the decrease of the orientation rate under the constant magnetic field would be the decrease in the anisotropy of magnetic susceptibility of the aligned material or the increase in the viscosity of the surrounding media.

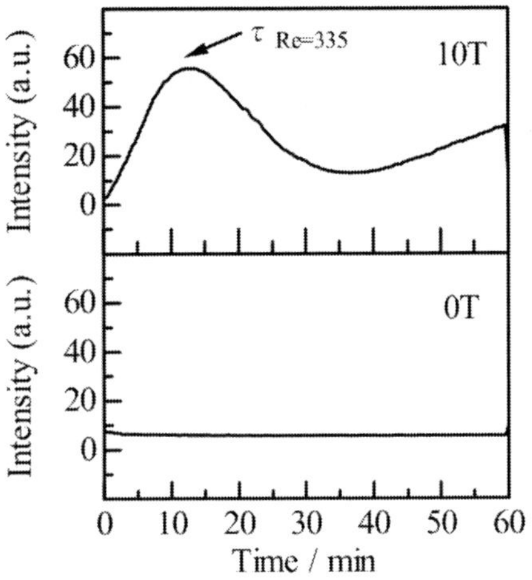

Figure 2. Change in the transmitting light intensity under cross-Nicole condition during annealing at 255 °C. $\tau_{\mathrm{Re}=335}$ is the time at which the intensity reaches the first maximum.

It is well known that the melting point determined by DSC measurement is an apparent melting point for crystalline polymers[10]. The equilibrium melting point of crystalline polymers is higher than that of the apparent melting point. It was reported that the equilibrium melting point of PET is 280 °C [11]. That is, the magnetic alignment occurring at 255 °C is regarded as occurring at a supercooling condition. In this situation, a structure formation such as a crystalline growth should progress. Though the viscosity increases with the structure formation, it is not considered that the anisotropy of the magnetic susceptibility is decreased. Hence, the decrease of the orientation rate originates from the increase of the viscosity due to the structure formation in a condition of lower supercooling.

We estimated the orientation rate to clarify the dependence of the orientation rate on the intensity of the magnetic flux density. If the retardation *Re* becomes 355 nm, the transmitting light intensity reaches the first maximum. We defined the time $\tau_{\mathrm{Re}=335}$ as the time at which the intensity maximum appears. The reciprocal of $\tau_{\mathrm{Re}=335}$ is proportional to the orientation rate which is expressed as $\tau^{-1} = F\chi_a B^2 / 6\mu_0\eta$, where F is the shape factor, χ_a is the anisotropic diamagnetic susceptibility of the material, B is the magnetic flux density, μ_0 is the magnetic permeability of vacuum, η is the viscosity of the surrounding media [1]. If the parameters governing the magnetic alignment are unchanged during the alignment, the orientation rate should be proportional to the square of the magnetic flux density. Figure 3 shows the relationship between the orientation rate and the square of the magnetic flux density. The orientation rate is estimated from the measurements of 2 or 3 times and the error bar is shown in the figure. The data points at lower magnetic fields greatly deviate from the straight line. This indicates that the orientation rate decreases at low magnetic fields in which the time for *Re* to reach 355 nm becomes longer. That is, the increasing in viscosity due to the structure formation should be remarkably responsible at lower magnetic fields. The effect of the increase in viscosity can not be ignored.

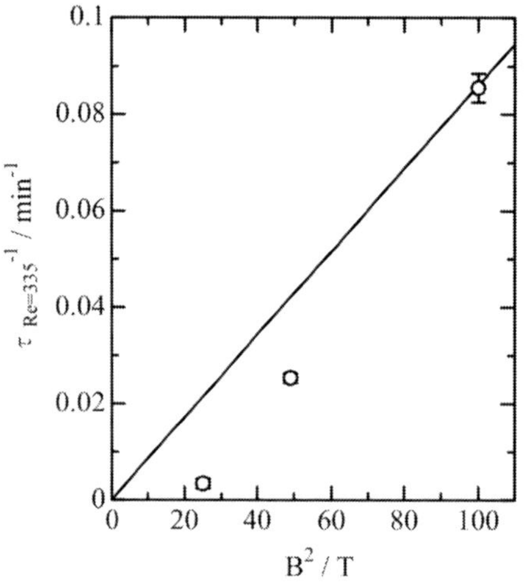

Figure 3. Relationship between the orientation rate and the square of the magnetic flux density.

Figure 4 shows the change in the transmitting light intensity under the cross-Nicole condition at different melting temperatures. With the increase of the melting temperature, the location of the intensity maximum shifts to longer annealing time and the transmitted light intensity disappeared at 260 °C. As mentioned above, the transmitting light intensity depends on Δn. Therefore, this result

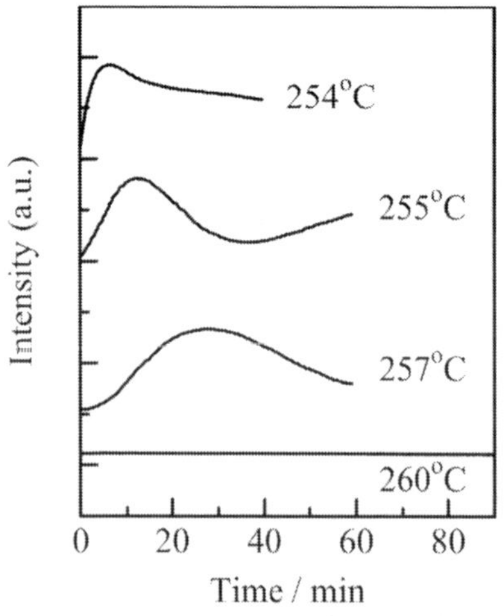

Figure 4. Change in the transmitting light intensity under cross-Nicole condition during annealing in a magnetic field (10T) at various annealing temperatures shown in the figure.

suggests that the annealing at higher temperatures causes the decrease of the increase rate of Δn.

Melting at higher temperature below the equilibrium melting point, the followings are expected.
1: The viscosity of polymer melt generally decreases with increase in the melting temperature.
2: The growth rate of the structure such as a crystal structure greatly decreases in the lower supercooling condition.
3: The number and size of the residual structure of crystallite decrease because the residual structure becomes more unstable.

If the viscosity decreases, the orientation rate should increase to result in the increase of the increase rate of Δn. Therefore, the decrease in the increase rate of Δn at higher melting temperatures is not attributed to the decrease of the orientation rate but to the decrease of the volume fraction of the ordered structures that can orient. The amount of the ordered structures is determined by two factors: the decrease of the residual ordered structures and the increase of newly grown ordered structures. At present time, it is difficult to distinguish which factor is dominant. This is our future work.

4. Conclusion

The magnetic alignment of poly(ethylene terephthalate) in molten state was studied by *in-situ* birefringence measurement in the magnetic field. We succeeded for the first time in detecting the magnetic alignment in the polymer melt by the *in-situ* observation of birefringence in a molten state at temperatures between 254 °C and 260 °C in the magnetic filed (10 T). Information of the structure formation in a low supercooling was obtained from the discussions about the change of the apparent orientation rate.

The dependence of the apparent orientation rate was due to the increase in the viscosity with the structure formation and the dependence of the apparent orientation rate on the melting temperature was due to the decrease of the volume fraction of the aligned structure accompanied with the decrease of the residual crystalline structure and/or the decrease in the growth rate of the structure.

ACKNOWLEDGEMENT

This work was partially supported by JSPS Asian Core Program "Construction of the World Center on Electromagnetic Processing of Materials".

References

[1] Kimura T 2003 *Polymer J.* **35** 823.
[2] Yamato M and Kimura T 2000 *Chem. Lett.* **2000** 1296.
[3] Sata H, Kimura T, Ogawa S, Yamato M and Ito E 1996 *Polymer* **37** 1879.
[4] Ezure H, Kimura T, Ogawa S and Ito E 1997 *Macromolecules* **30** 3600.
[5] Kawai T and Kimura T 2000 *Polymer* **41** 155.
[6] Kimura T, Kawai T and Sakamoto Y 2000 *Polymer* **41** 809.
[7] Aoki H, Yamato M and Kimura T 2001 *Chem. Lett.* **2001** 1140.
[8] Yamato M and Kimura T 2002 *Trans MRS-J* **27** 117.
[9] Ebert F and Thurn-Albrecht T 2003 *Macromolecule* **36** 8685.
[10] Wunderlich B 1976 *Macromolecular Physics* vol 2 (New York: Academic Press).
[11] Rober R C 1969 *Polymer* **10** 114.

Journal of Physics: Conference Series **156** (2009) 012002 doi:10.1088/1742-6596/156/1/012002

Three-dimensional orientation of cellulose crystals under dynamic elliptic magnetic field

F Kimura and T Kimura

Division of Forest and Biomaterials Science, Graduate School of Agriculture, Kyoto University, Sakyo-ku, Kyoto 606-8502, Japan

E-mail: fkimura@kais.kyoto-u.ac.jp

Abstract. Cellulose microparticles prepared by hydrochloric acid hydrolysis of cotton cellulose were mixed with UV-curable resin precursor and subjected to a frequency modulated magnetic field. The obtained alignment was fixed by photopolymerization of the matrix resin precursor. The observation of the obtained sample from three different orthogonal directions exhibited optical anisotropy, indicating that the three-dimensional alignment is achieved.

1. Introduction

Recently, we have developed a new technique that enables three-dimensional alignment of crystalline powders [1,2]. This technique is based on the use of dynamically modulated (elliptic) magnetic fields. With this technique, we can prepare from a powder sample a pseudo single crystal in which the individual crystallites are aligned three-dimensionally. This technique is applicable to the biaxial crystal systems including the orthorhombic, monoclinic, and triclinic systems. An excellent property of the pseudo single crystal is that it gives rise to the X-ray diffraction pattern identical with that obtained from a real single crystal [2], enabling a single crystal analysis. This technique is also useful for the processing of composites in which crystalline fillers are aligned three-dimensionally.

It is known that the single crystal analysis with X-ray diffraction is the most useful technique for determination of the three dimensional structure of new compounds. However, growing a single crystal to a size suitable for the X-ray measurement is becoming more difficult because the structure of the compound is getting more complicated. Recently, new analytical techniques such as the *ab-initio* or direct space method have been proposed [3-5] to solve the crystal structure from a powder diffraction pattern. Unfortunately, these techniques are limited to specialists because the determination of the correct cell constant and the structure is not easy from one-dimensional information. A pseudo single crystal can provide sufficient data for the determination of the crystal structure.

In this paper, we report a preparation of a pseudo single crystal of cellulose microcrystals. Natural cellulose fibers are of polycrystal of microcrystals, so the decomposition to the microcrystal is a key issue. In addition, microcrystals from naturally occurring cellulose have two crystal forms: I_α (triclinic) and I_β forms (monoclinic). A microcrystal is a composite of the two forms, the ratio of which depends on sources. At present time, we do not have enough knowledge regarding to which size a polycrystal should be decomposed in order to obtain microcrystals. In this paper, we report a preliminary work on the preparation of a pseudo single crystal.

Journal of Physics: Conference Series **156** (2009) 012002 doi:10.1088/1742-6596/156/1/012002

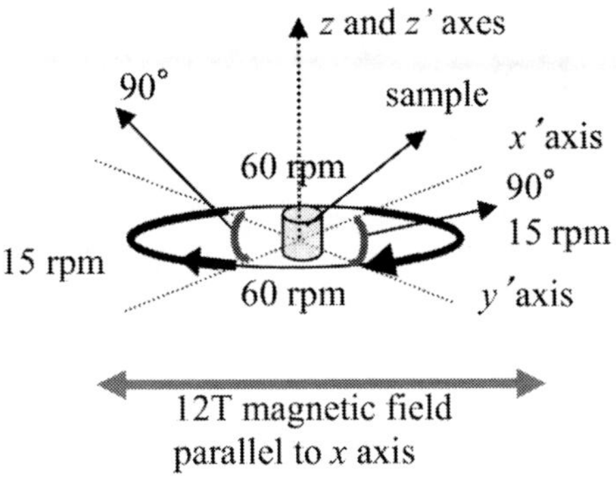

Figure 1. Diagram illustrating the frequency-modulated sample rotation. The x, y, and z axes are laboratory coordinates; the x and y axes are on the horizontal plane and the z axis is in the vertical direction. The x', y', and z' axes are imbedded in the sample that is rotated around the z' axis (parallel to the z axis). The rotation is performed nonuniformly. The x axis coincides with the direction of the magnetic field.

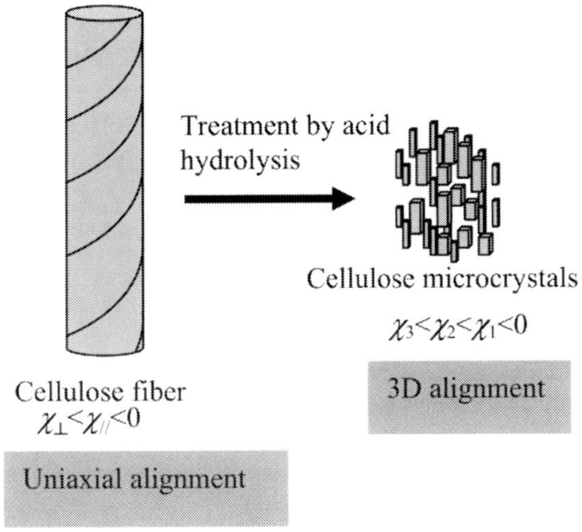

Figure 2. Magnetic anisotropy of cellulose fiber and microcrystal. $\chi_{//}$ and $\chi_\perp$ are magnetic susceptibilities parallel and perpendicular to fiber axis, respectively. χ_1, χ_2, and χ_3 are three different susceptibilities of microcrystal. A polycrystal (a fiber) is decomposed to single crystals (microcrystals) by acid hydrolysis. A fiber is uniaxial while a microcrystal (triclinic (I_α form) or monoclinic (I_β)) is biaxial.

2. Experimentals

Whatman CF11 cellulose (cotton cellulose) was mixed with 4 N hydrochloric acid and stirred at 80 °C for 4 h to obtain a cellulose suspension. The acid was removed by centrifugation and prolonged dialysis with pure water. An ethanol solution of saturated sodium dodecyl sulfonate (0.5ml) was added to the cellulose suspension (ca. 40ml). Supernatant of the suspension was removed by decantation, and the ethanol was added to the remaining suspension; this procedure was repeated several times to replace the water with ethanol. The ethanol was evaporated at room temperature until the sample became paste-like. Then a UV-curable resin precursor (No. 8815 of Kyoritsu Chemical and Co. Ltd., viscosity of 1200 mPa s) was added to the paste to obtain a suspension with resin precursor. A small amount of the resin suspension was poured into a plastic container with 5 mm diameter and 10 mm height. The container was mounted on a sample-rotating unit inside a superconducting magnet (12 T). Two hours after exposure to the frequency-modulated sample rotation, the resin was irradiated with UV light inside the magnet for 30 minutes to fix the alignment.

The sample rotating unit was placed in the center of the bore of a superconducting magnet generating a 12-T horizontal magnetic field (the x axis). The rotation axis was vertical (the z axis). The rotation was not uniform, but the frequency was periodically changed every 90 degree. In the present study, we set the values of the frequency as 15 and 60 rpm (figure 1).

3. Results and discussion

Under a static magnetic field, the easy magnetization axis aligns parallel to the applied field, while the hard magnetization axis has a freedom of rotation around the field direction, resulting in a planar alignment on the plane perpendicular to the applied field. Under a rotating magnetic field, the hard magnetization axis aligns perpendicular to the plane of rotation of the field, while the direction of the easy magnetization axis is not fixed. That is, the hard axis aligns uniaxially. By using a dynamic modulated elliptic magnetic field, both the easy and hard axes are fixed with respect to the sample coordinates. Although a cellulose microcrystal has a biaxial

Figure 3. Polarized microscope observation (with a color plate) of three dimensional alignment (a pseudo single crystal) of cellulose microcrystals achieved by sample modulated rotation. Red arrows and dots indicate the direction of the rotating axis that corresponds to the hard magnetization axis. Top view (a), side view 1 (b) and side view 2 (c) clearly indicate the three dimensional alignment.

nature, a large cellulose fiber only exhibits the uniaxial nature because it is polycrystal. Therefore, it is necessary to decompose a large fiber to microcrystals in order to achieve three-dimensional alignment (figure 2). If the decomposition of large fibers is made by an appropriate acid treatment, we will obtain microcrystals that undergo three-dimensional alignment. The decomposed sample mainly exhibited the I_β form (monoclinic) according to the X-ray diffraction analysis (results are not shown).

Figure 3 shows the optical polarized microscopic photographs with a color plate of the cellulose microparticles/resin composite obtained from the application of modulated magnetic field. The images in figure 3(a) (through view) were taken from the direction of the hard magnetization axis (the fiber axis). If the decomposed particles are yet polycrystal (that is, unixaial with respect to the fiber axis), we do not observed color change when the sample was rotated 90 degree under the polarized microscope. However, we observed the color change. Upper images in figures 3 (b) and (c) (side-views 1 and 2) showed blue and orange colors even if the tilt angles of the sample are the same. If the particles were yet polycrystal, we would observe the same color at the same tilt angles. These observations indicate that the decomposes particles are not polycrystals but micro (single) crystals. From these results, we conclude that we succeeded to obtain a pseudo single crystal of cellulose I_β form. Unfortunately, the X-ray diffraction measurement was not successful because the concentration of the microcrystal was not enough high to obtain diffractions.

References

[1] Kimura T and Yoshino Y 2005 *Langmuir* **21** 4805-8
[2] Kimura T, Kimura F, and Yoshino M 2006 *Langmuir* **22** 3464-6
[3] Harris K. D M, and Tremayne M 1996 *Chem. Mater.* **8** 2554-70
[4] David W I F, Shankland K, and Shankland N 1998 *Chem. Commun.* **1998** 931-2
[5] Wu J, Leinenweber K, Spence J C H, and O'Keeffe M 2006 *Nature Materials* **5** 647-52

Continuous Measurement of XRD Peak Intensity of Bismuth under Magnetic Field

T Kohama and K Iwai

Department of Material, Physics and Energy Engineering, Nagoya University, Nagoya 464-8603, Japan

E-mail: kohama.takenori@g.mbox.nagoya-u.ac.jp

Abstract. A material which has a magnetic anisotropy rotates to reduce magnetization energy under a magnetic field, and finally the magnetic easy axis and the direction of magnetic field becomes the same direction. For clarification of the crystal rotation, crystal alignment benaviour of bismuth which has a magnetically anisotropic nature was examined in situ by X-ray diffraction during the static magnetic field imposition. Change of the bismuth (110) peak intensity in time was measured by X-ray diffraction to evaluate the crystal alignment. The alignment time decreased as the magnetic field strength increased. This tendency was similar to the theoretically calculated relaxation time. Aggregation of the bismuth particles decreases the driving force for the crystal alignment, which was in proportion to the magnetic susceptibility difference between the magnetically easy and hard axes. The effective difference in magnetic susceptibility for aggregated bismuth particles was estimated by measuring the alignment time of the particles under magnetic fields of various strengths. The estimated effective magnetic susceptibility difference increases with a decreasing magnetic field strength.

1. Introduction

Imposition of a magnetic field on materials has been utilized in materials processes such as crystal alignment [1], levitation [2], separation [3], and flow control of an electrically conductive material. In particular, alignment process is an attractive tool for producing functional materials, because physical properties of materials, such as the electric, magnetic, thermal, and mechanical properties, can be controlled if these properties have an anisotropic nature. With the development of the technology of superconducting magnets, crystals of nonmagnetic, magnetically anisotropic materials such as metals [4-6], ceramics [7, 8], and polymers [9] can be aligned by the imposition of a strong magnetic field. To form unidirectionally aligned crystals for a material with a magnetic anisotropy of $\chi_c < \chi_a$, a combined process involving the imposition of a magnetic field and the rotation of a sample has been proposed [9, 10]. By using this method, a bulk form of hydroxyapatite with unidirectionally aligned crystals has been produced [11]. A theoretical analysis for the optimization of this process has been performed [12]. Furthermore, crystal alignment of a tin–lead binary alloy primary phase in a magnetically preferred direction can be controlled by changing the duration of an imposed electric current and static magnetic field during solidification [13]. For optimization of these crystal-alignment processes, the rotation behaviour of crystals suspended in a solvent in a magnetic field should be clarified. Therefore, alignment behaviour of feeble magnetic material has been observed by using a laser microscope [14] and theoretical development of crystal alignment behaviour has been done [15].

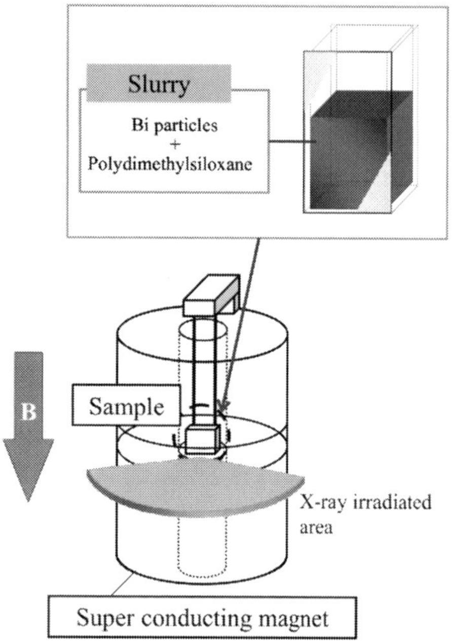

Figure 1. Experimental apparatus

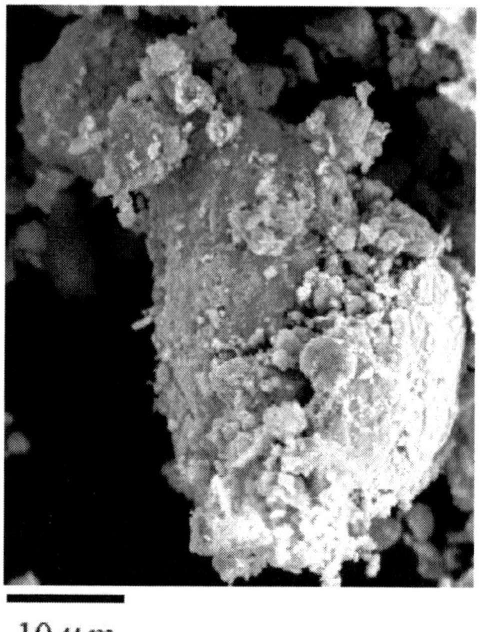

Figure 2. SEM picture of bismuth particle

In this paper, we examined the in situ crystal-alignment behaviour of bismuth particles in the presence of a static magnetic field by in situ x-ray diffraction (XRD) studies.

2. Experimental

The experimental apparatus is shown in Fig.1. Bismuth particles (2.67 g) mixed with liquid polydimethylsiloxane (4 mL) was used as a slurry in this research. The viscosity and density of the polydimethylsiloxane were 50 Pa.s and 1000 kg/m^3, respectively. An acrylic vessel containing the slurry was placed in the bore of a superconducting magnet for the imposition a static magnetic field on the slurry. The direction of magnetic field was vertical downward direction.

The crystallographic structure of bismuth can be considered as a hexagonal structure and bismuth crystals show anisotropy in magnetic susceptibility. The magnetic susceptibility of bismuth in the direction of the a-axis is -1.77×10^{-4} and that in the direction of the c-axis is -1.24×10^{-4}, so that c-axis is parallel to the magnetic field direction if a magnetic field is imposed on the bismuth crystal. The X-ray direction in this experiment was in horizontal plane, so the XRD peak intensity in a plane parallel to the c-axis increased when the magnetic field was imposed on bismuth crystal.

The difference of magnetic susceptibility of a single crystal in easy axis and hard axis is described as "magnetic susceptibility difference", and that for an aggregated particle is described as "effective magnetic susceptibility difference" in this manuscript. Magnetic susceptibility difference and effective magnetic susceptibility difference are expressed as $\Delta\chi$ and $\Delta\chi_E$, respectively.

The effective magnetic susceptibility difference of the bismuth particles used in this experiment was small in comparison with the magnetic susceptibility difference of a single crystal, because the bismuth particles with non-spherical shapes aggregated as shown in Fig.2. Size distribution of the bismuth particles used in this experiment was from 0.259 μm to 34 μm and the mean particle size was 8.25 μm, respectively.

The bismuth particles moves by gravitational and magnetization forces under the magnetic field. These two forces acting on the particles were evaluated. The driving force for sedimentation of the bismuth particle is the difference in density between the bismuth particles and polydimethylsiloxane, which can be expressed as follows.

$$F_g = (\rho_{Bi} - \rho_{polymer}) g V_{Bi} \tag{1}$$

where g is the acceleration of gravity, V_{Bi} is the volume of bismuth particles, ρ_{Bi} is the density of bismuth, and $\rho_{polymer}$ is the density of polydimethylsiloxane, respectively.

This force is calculated as 2.31×10^{-11} N for a spherical particle with a diameter of 8.25 μ m.

In the presence of the magnetic field, the magnetization force affects the motion of the bismuth particles. The following equations are used to evaluate the magnetization force.

$$\begin{cases} F_r = \mu_0 (\chi_{Bi} - \chi_{polymer}) \left(H_r \frac{\partial}{\partial r} H_r + H_z \frac{\partial}{\partial z} H_r \right) V_{Bi} \\ F_z = \mu_0 (\chi_{Bi} - \chi_{polymer}) \left(H_r \frac{\partial}{\partial r} H_z + H_z \frac{\partial}{\partial z} H_z \right) V_{Bi} \end{cases} \tag{2}$$

where r is the radial direction, z is the axial direction, μ_0 is the magnetic permeability in vacuum, and χ_{Bi} and $\chi_{polymer}$ are magnetic susceptibilities of bismuth and polydimethylsiloxane, respectively.

The magnetization force is therefore proportional to the difference in magnetic susceptibility between the bismuth particles and the polydimethylsiloxane. The magnetic susceptibility of the polydimethylsiloxane is neglected in the calculation, because the magnetic susceptibility of polymers is usually of the order of 10^{-6} [16] which is small in comparison with that of bismuth. The calculated forces for a 8.25 μ m-diameter spherical bismuth particle in the radial and axial directions are Fr = 1.75×10^{-13} N and Fz = 9.73×10^{-13} N, respectively, when the strength of the magnetic field is 3 T. Therefore, main force acting on the bismuth particles is the gravitational force.

Brownian motion acts on the bismuth particles as a thermal fluctuation. The relaxation time for the Brownian motion, τ_B is expressed as follows [17].

$$\tau_B = \frac{3 V_{Bi} \eta}{kT} \tag{3}$$

where η is viscosity, k is Boltzmann constant and T is absolute temperature, respectively.

On the other hand, the temporal variation of the angle θ for the particle rotation by the magnetic torque is theoretically expressed as follows when the inertial term can be neglected [18].

$$\tan\theta = \tan\theta_0 \exp\left(-\frac{t}{\tau} \right) \tag{4}$$

$$\tau = \frac{30\eta + R^2 \sigma \mu_0^2 H^2}{5 \Delta\chi_E \mu_0^2 H^2} \mu_0 \tag{5}$$

where θ_0 is the initial angle between the magnetic field direction and the magnetically easy axis and τ is the relaxation time for the crystal alignment, respectively, and H, R, t, θ, and σ are the magnetic field strength, crystal radius, time, the angle between the magnetically easy axis of the crystal and the magnetic field direction and the electric conductivity, respectively.

When the relaxation time for the Brownian motion, τ_B is equal to the relaxation time for the crystal alignment τ, the critical diameter for a spherical particle, d_c can be obtained as follows under the assumption that the effect of the electrical conductivity can be neglected.

$$d_c = 2 \left(\frac{3 \mu_0 k T}{2 \pi \Delta\chi_E B^2} \right)^{\frac{1}{3}} \tag{6}$$

The critical particle sizes in the cases of B=0.5 T and B=1 T magnetic fields are calculated to 0.246 μm and 0.155 μm, respectively when the effective magnetic susceptibility difference $\Delta\chi_E$ is 5.3 $\times 10^{-6}$ and T is 300 K. The effective magnetic susceptibility adopted here is one-tenth of the single crystal value. Therefore Brownian motion might slightly affect the alignment behavior under these

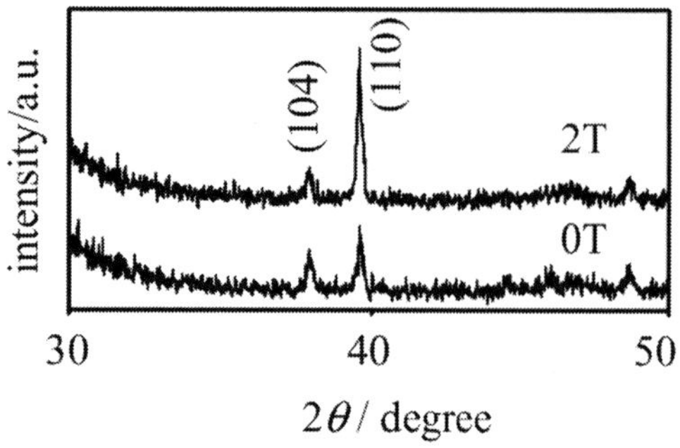

Figure 3. XRD results with and without magnetic field

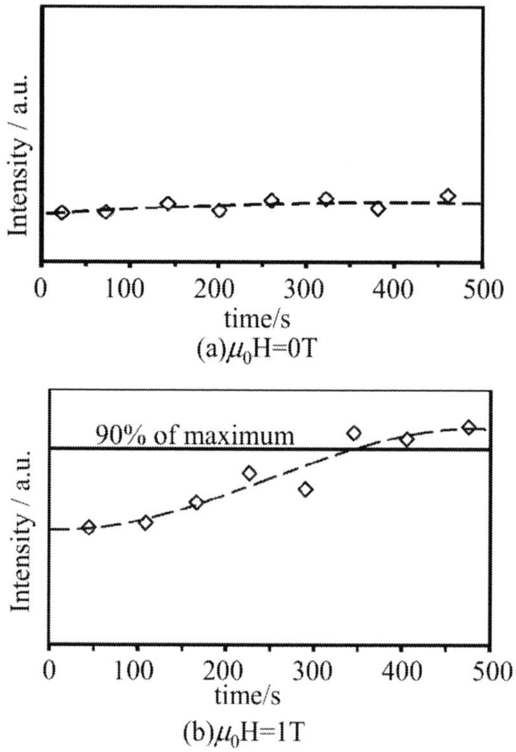

Figure 4. Temporal change of Bi (110)

Table 1. Alignmemt time under different mangnetic field strength

μ_0H / T	Alignment time / s
0.5	330
1	330
1.5	135
2	70
3	85

magnetic fields because the minimum size of the bismuth particles used in this experiment was 0.259 μm.

The XRD results with and without a 2 T magnetic field are shown in Fig. 3. The (110) peak increased by imposing a magnetic field, whereas the (104) peak relatively decreased. This well agrees with the theoretical prediction mentioned above while aggregated bismuth particles were used in the experiment. Therefore, the (110) peak intensity of the bismuth crystal was chosen as an index of the crystal alignment. And its temporal change was measured for various strengths of the magnetic field. The results are shown in Fig.4 with a fitting curve. The (110) peak intensity did not change with time under the no magnetic field although it showed a small fluctuation. This means that sedimentation of the bismuth particles in this experiment can be neglected. And the magnetization forces in radial and axial directions can therefore be ignored in this experiment, since these forces are smaller than the force caused by the density difference. The fluctuations observed here are might be caused by thermal fluctuation. On the other hand, the intensity of the (110) peak measured under a 1 T magnetic field increased gradually with time and it reached a plateau after a certain time. That is, the bismuth particles in the polydimethylsiloxane rotate to the magnetically preferred direction. We defined the alignment time in this experiment as the period from the insertion of the sample into the bore of the magnet to the time at which the (110) peak intensity reached 90% of its maximum value. The 90% value is also indicated in Fig.4(b) as the horizontal line.

Table1 shows the alignment time in the several different magnetic field strengths. The alignment time decreases with increasing magnetic field strength. In particular, it decreases markedly between the 1 T and 2 T. The alignment time at fields of more than 2 T may be weakly dependent on the strength of the magnetic field, but this is not clear from this experiment.

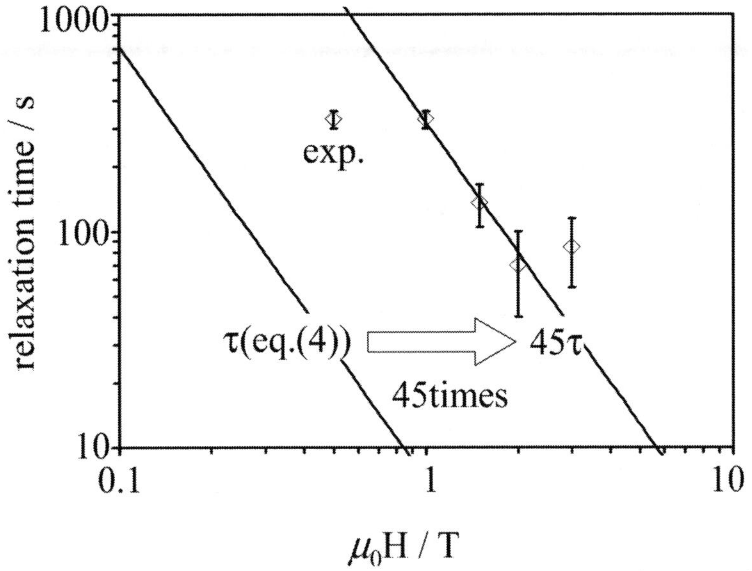

Figure 5. Comparison of calculated alignment time and measured one

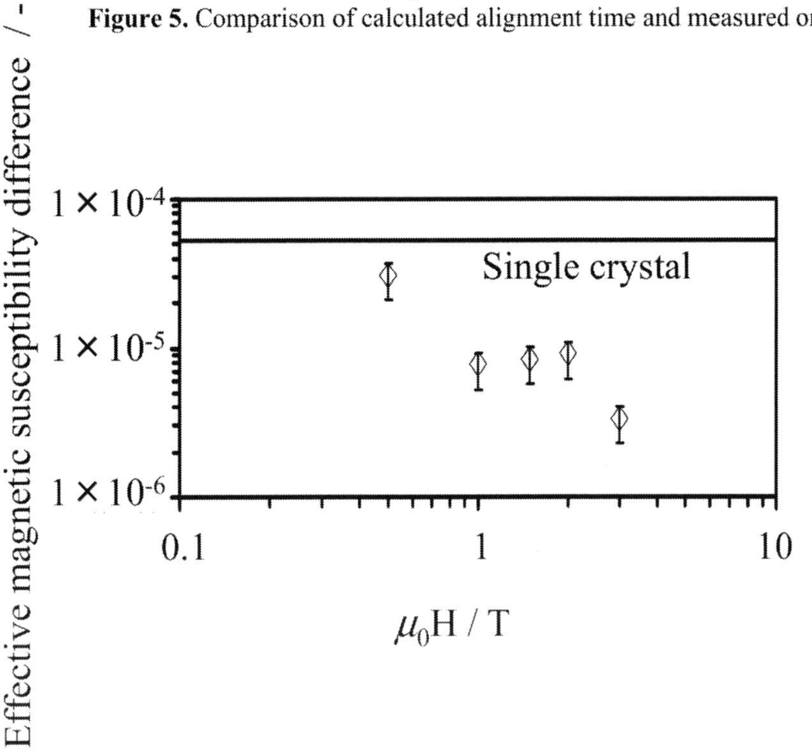

Figure 6. Estimated effective magnetic susceptibility difference of the bismuth particles

3. Discussion

The relaxation time was calculated by using equation (4) and was compared with the measured alignment time. The results are shown in Fig.5. The real alignment time exists between two measurement times whose interval is about 30 seconds as shown in the Fig.4. Therefore this interval is adopted as the range of error bar in Fig. 5. The relation between the magnetic field strength and the measured alignment time is similar to the relation between the magnetic field strength and the

Table 2. Effect of initial angle on effective magnetic susceptibility difference

Initial angle θ_0 / degree	0.5T	1T	1.5T	2T	3T
1	1.1×10^{-5}	2.6×10^{-6}	2.9×10^{-6}	3.1×10^{-6}	1.1×10^{-6}
10	2.1×10^{-5}	5.3×10^{-6}	5.7×10^{-6}	6.2×10^{-6}	2.3×10^{-6}
57.3	3.1×10^{-5}	7.8×10^{-6}	8.4×10^{-6}	9.1×10^{-6}	3.3×10^{-6}
80	3.7×10^{-5}	9.2×10^{-6}	1.0×10^{-5}	1.1×10^{-5}	4.0×10^{-6}
89	4.8×10^{-5}	1.2×10^{-5}	1.3×10^{-5}	1.4×10^{-5}	5.1×10^{-6}

calculated relaxation time, although the measured alignment time is longer than the calculated relaxation time under the same strength of magnetic field. This is mainly because the effective magnetic susceptibility difference in the experiment is small in comparison with that of single crystal. Actually, the forty five times increase in the value of the relaxation time agrees well with the measured alignment time except the result under the 0.5T magnetic field, as shown in Fig.5.

The average angle between the direction of the magnetic field and the c-axis, θ_i immediately after the imposition of the magnetic field was theoretically calculated to one radian based on the random distribution of the particles. The effective magnetic susceptibility difference, $\Delta\chi_E$ was then estimated from equations (3) and (4), by assuming that the final and initial angles were 0.1° (0.0017 radians) and 57.3° (1 radian), respectively. According to the final angle, it should be theoretically decided under the consideration of thermal fluctuation while the 0.1° was adopted here. The alignment times shown in Table1 were also used in the calculation. Figure 6 shows the calculated effective magnetic susceptibility difference of the bismuth particles under the different magnetic field strength.

Horizontal line in Fig.6 is the magnetic susceptibility difference of bismuth single crystal, $\Delta\chi$. The estimated effective magnetic susceptibility difference, $\Delta\chi_E$ has roughly a tendency to decrease with the increase in the magnetic field strength. This is because only particles with a large value in effective magnetic susceptibility align in the magnetically preferred direction under a weak magnetic field, whereas particles with not only a large value in effective magnetic susceptibility difference but also a small value in effective magnetic susceptibility difference also align in the magnetically preferred direction under a strong magnetic field.

However, slight increase in the effective magnetic susceptibility measured between 1T and 2T can not be explained only from the viewpoint of the measuring time error.

To confirm the accuracy of the evaluated effective magnetic susceptibility, we calculated the effective magnetic susceptibility difference under the different initial angles of 1° (0.017 radians), 10° (0.17 radians), 57.3° (1 radian), 80° (1.4radians) and 89° (1.6 radians). The final angle was fixed to 0.1 ° (0.0017 radians). Table2 shows the calculated values of the effective magnetic susceptibility difference under different initial angles. The effective magnetic susceptibility difference weakly depends on the initial angle. The error due to the initial angle difference was estimated within 43% if it was between 10° (0.17 radians) and 80° (1.4 radians) in this experiment. The error bars in Fig.6 are evaluated by taking account of fluctuation of initial angle, whose minimum and maximum valued are assumed to be 10° and 80° , respectively. The error induced by the measuring time is small in comparison the error induced by the initial angle in every experimental condition. For more accurate measurement of the effective magnetic susceptibility difference, we should align the particles to a certain angle in the experiment. This is our future work.

4. Conclusion

By using XRD analysis, we examined the alignment behaviour of bismuth particles in order to elucidate their rotational behaviour when suspended in a solvent in a static magnetic field. The following is the main result that we obtained.

The effective magnetic susceptibility difference of aggregated particles, $\Delta\chi_E$ can be estimated by measuring the alignment time of the particles in a magnetic field.
Particles with a small effective magnetic susceptibility difference rotate under strong magnetic field

Acknowledgement
This research was partially supported by Iron and Steel Institute of Japan and JSPS Asian Core Program "Construction of the World Center on Electromagnetic Processing of Materials".

References

[1] Li S, Sassa K, Iwai K and Asai S 2004 *Mater. Trans.* **45** 3132-3129
[2] Iwai K and Asai S 1995 *Modeling and Simulation in Materials Science and Engineering* **3** 473-484
[3] Tanaka Y, Sassa K, Iwai K and S.Asai 1995 *Tetsu to Hagane* **81** 1120-1125
[4] Morikawa H, Sassa K and Asai S 1998 *Mater. Trans, JIM.* **39** 814-818
[5] Tahashi M, Ishihara M, Sassa K and Asai S 2003 *Mater. Trans, JIM.* **44** 285-289
[6] Usui M, Iwai K and Asai S 2006 *ISIJ Int.* **46** 859-863
[7] Suzuki T, Sakka Y and Kitazawa K 2001 *J.Cera.Soc.JP.* **109** 886-890
[8] Sakka Y and Suzuki S 2005 *J.Cera.Soc.JP.* **113** 26-36
[9] Kimura T 2003 *Polymer Journal* **35** 823-843
[10] Akiyama J, Hashimoto M, Takadama H, Nagata F, Yokogawa Y, Sassa K, Iwai K and Asai S 2005 *Mater. Trans.* **46** 203-206
[11] Akiyama J, Hashimoto M, Takadama H, Nagata F, Yokogawa Y, Sassa K, Iwai K and Asai S 2005 *Mater. Trans.* **46** 2514-1517
[12] Akiyama J, Asano H, Iwai K and Asai S 2007 *Mater. Trans. JIM* **71** 108-112
[13] Usui M, Iwai K and Asai S 2006 *ISIJ Int.* **46** 859-863
[14] Hirota N, Sakka Y, Shimada T, Ando T and Wada H 2007 *The Magnetic-Science Society of Japan 2nd Annual Meeting Program and Abstract* 1O-04
[15] Yamaguchi M, Yamamoto I and Ogawa K 2007 *The Magnetic-Science Society of Japan 2nd Annual Meeting Program and Abstract* 1O-06
[16] Yamato M 2004 *Kobunshi Ronbunshu.* **61** 433-441
[17] Wu Cunyou, Li Shuqin, Sassa Kensuke, Chino Yasumasa, Hattori Kazutoshi and Asai Shigeo 2005 *Mater. Trans.* **46** 1311-1317
[18] Sugiyama T, Tahashi M, Sassa K and Asai S 2003 *ISIJ Int.* **43** 855-861

Journal of Physics: Conference Series **156** (2009) 012004 doi:10.1088/1742-6596/156/1/012004

Different properties exhibited on the two typical crystal faces of hydroxyapatite in a simulated body environment

T Hagio[1], T Tanase[2], J Akiyama[3], K Iwai[1] and S Asai[4]

[1]Department of Materials, Physics and Energy Engineering, Nagoya University, Furo-cho, Chikusa-ku, Nagoya, 464-8603, Japan

[2]Toho Gas Corporation, 19-18 Sakurada-cho, Atsuta-ku, Nagoya, 456-8511, Japan

[3]Institute for Molecular Science, National Institute of Natural Sciences, 38 Nishigonaka, Myodaiji, Okazaki, 444-8585, Japan

[4]Innovation Plaza Tokai Japan Science and Technology Agency, 23-1 Ahara-cho, Minami-ku, Nagoya, 457-0063, Japan

E-mail: hagio.takeshi@h.mbox.nagoya-u.ac.jp

Abstract. Hydroxyapatite is a main mineral constituent of hard tissues and is extensively used as a biomaterial in the medical field. Hydroxyapatite exhibits anisotropic chemical properties on its two typical crystal faces, the a-face and c-face, due to its hexagonal crystal structure. In polycrystalline bodies, such anisotropy can be enhanced by controlling the crystal orientation. Hydroxyapatite nanocrystals that constitute long bones also form a crystal oriented structure. Therefore, clarification of the difference between the properties of a-face and c-face in hydroxyapatite using in vitro experiments is useful to understand the structure and function of actual hard tissues. Hydroxyapatite ceramics with controlled crystal orientation were prepared by a slip casting method under application of a magnetic field. The fabricated hydroxyapatite ceramics had surfaces consisting mainly of a-face, or otherwise c-face hydroxyapatite. These hydroxyapatite ceramics were immersed into simulated body fluids to investigate the difference in bioactivity. The precipitation behavior observed on the surface of each hydroxyapatite ceramic was different. The thickness of the precipitate was increased and formed earlier on the HAp c-face compared to that on the a-face.

1. Introduction

In line with the aging population of Japan, there has been much emphasis on the development of biomedical materials for tissue engineering. One of the most typical biomedical materials extensively applied in the biomedical field is hydroxyapatite (HAp: $Ca_{10}(PO_4)_6(OH)_2$) [1-3], a type of calcium phosphate that comprises a substantial portion of the mineral constituent found in human hard tissues [4-9]. Therefore, HAp has received much attention from many researchers [6-8] and is considered to be a promising biomaterial with high biological affinity [8-9].

HAp is also known for its anisotropic properties; as a result of its hexagonal crystal structure, HAp exhibits anisotropic chemical properties depending on its two typical crystal faces, the a-face and c-face. One such example is the anisotropy in adsorption, and much research has been reported on the crystal face dependent adsorption properties of Hap [8, 10-15]. This adsorption dependency is due to the difference in the dominant atoms composing each crystal face of HAp; positively charged calcium on the a-face and negatively charged phosphate groups on the c-face [14-15]. For example, Kawachi et al. [8] reported that porous polycrystalline apatite with a larger a-face surface area than c-face had a higher selectivity for adsorption of bovine serum albumin (acidic protein) than lysozyme chloride (basic protein).

In polycrystalline bodies, such anisotropic properties can be enhanced by controlling the crystal orientation. Magnetic field is a useful tool to control the crystal orientation of magnetically anisotropic materials [16-25]. Due to the recent developments in superconducting technology, magnetic orientation has been extended to diamagnetic materials such as polymers [18] and ceramics [19-25]. For example, Kimura et al. [18] reported the magnetic orientation of various polymers, and Tanaka et al. [20] reported that the c-axis of ZnO can be oriented using a rotating magnetic field with subsequent sintering. The magnetic orientation of HAp was first reported by Inoue et al. [21]. Sakka et al. [22,23] disclosed the effect of agglomeration on the texture of HAp ceramics fabricated by slip casting under a high magnetic field, and concluded that highly textured HAp could be fabricated by slip casting with a well dispersed suspension of HAp single crystals in a high magnetic field followed by sintering. Furthermore, Akiyama et al. [24,25] reported that polycrystalline HAp with c-axis orientation could be formed by slip casting under a high magnetic field with mechanical sample rotation.

The hard tissues in humans are composed of HAp nanocrystals that form a self-organized structure exhibiting crystal orientation in a particular direction [26]. Taking the human long bone for example, the c-axis of HAp is oriented in the direction parallel to the ultimate load direction [26]. The crystal orientation enhances the anisotropic properties of HAp; therefore, the long bone selectively utilizes the anisotropic properties that depend on the two typical crystal faces of HAp. However, the anisotropy in bioactivity exhibited on these two crystal faces in a simulated body environment has not been investigated until now.

Simulated body fluid (SBF) is a valid medium for evaluation of the bioactivity of such a material, and much research using SBF has been reported to date[27-34]. Kokubo et al. [31] reviewed the in vitro ability of various materials to form apatite layers on their surfaces in SBF and their corresponding in vivo bioactivities, and concluded that examination of apatite formation on a material in SBF is useful for predicting the in vivo bioactivity of a material. They also reported that a material that is able to form apatite on its surface in SBF can bond to living bone in a shorter period. Yamashita et al. [32] reported that the bone-like crystal growth on polarized HAp ceramics immersed in supersaturated SBF (1.5SBF) was accelerated on a negatively charged surface and decelerated on a positively charged surface. Tanahashi et al.[33] have investigated the surface functional group dependence on apatite formation on self-assembled monolayers (SAMs) in SBF and reported that SAMs of alkanethiols with negatively charged groups, such as phosphoric (-PO4H2) and carboxyl groups (-COOH), strongly induced apatite formation, while the positively charged amino group (-NH2) did not. Kim et al.[34] reported that the initial bone-like apatite formation took place earlier on the surface of HAp sintered at a lower temperature, which had a stronger negative charge, by attracting calcium ions to form calcium-rich amorphous or nano-crystalline calcium phosphate on its surface in SBF.

In this study, HAp ceramics with controlled orientation have been prepared by a slip casting method under the imposition of a magnetic field. The fabricated HAp ceramics had a surface mainly consisting of a-face or otherwise c-face of HAp, and in vitro experiment were carried out to examine the difference in bioactivity between the two typical crystal faces of HAp in a simulated body environment provided by SBF.

2. Experimental Procedure

2.1. Preparation of Crystal Oriented HAp Ceramics

Table 1. Ion Concentration of c-SBF

Positive ion	Concentration [mol/m^3]	Negative ion	Concentration [mol/m^3]
Na^+	142.0	Cl^-	147.8
K^+	5.0	HCO_3^-	4.2
Mg^{2+}	1.5	HPO_4^{2-}	1.0
Ca^{2+}	2.5	SO_4^{2-}	0.5

HAp particles (HAp-200, Taihei Chemical Co.) were dispersed and suspended in ion-exchanged water using a dispersing agent (CELUNA D-305; Chukyo Yushi Co.). The mixture was mechanically ground in an agate mortar using a milling machine (Type ANM1000, Nitto Kagaku Co. Ltd.) to break down aggregated particles. After milling for over 4 hours, the slurry was poured into gypsum molds and slip cast, with or without the application of a 10 T magnetic field, to fabricate HAp green compacts with controlled orientation or with random orientation, respectively. In the former case, a green compact with a-face orientation was fabricated using a vertical magnetic field [22,23], while that with c-face orientation was fabricated using a horizontal magnetic field with mechanical mold rotation [24,25]. The green compacts were dried in air at room temperature for over 24 hours in the absence of a magnetic field, and were the sintered at 1573 K in an electric furnace under atmospheric pressure. The sintered HAp ceramics were discoid in shape with diameters and thicknesses of approximately 12 mm and 2 mm, respectively. The crystal orientation and grain size distribution of the three HAp ceramics were then evaluated.

2.2. Evaluation of Crystal Orientation and Grain Size Distribution

The crystal orientation of the HAp ceramics was evaluated using X-ray diffraction (XRD; cat. no. 2035G403, Rigaku Corp.) with Cu-Kα radiation, after the ceramics surfaces were polished with emery paper of grade #2000. Scanning electron microscopy (SEM; VE-7800, Keyence Corp.) was used to evaluate the grain size distribution by examination of 82 to 104 grains in SEM images of each sample etched with HNO_3.

2.3. Evaluation of Bioactivity using SBF

Conventional SBF (c-SBF) was prepared by dissolving NaCl, $NaHCO_3$, KCl, K_2HPO_4, $MgCl_2 \cdot 6H_2O$, $CaCl_2$, Na_2SO_4, $(CH_2OH)_3(CNH)_2$ (buffer) and HCl (pH adjuster) in pure water, as proposed by Kokubo et al.[27-31]. The ion concentration of c-SBF is shown in Table 1.

The HAp ceramics were polished with 0.05 μm alumina buff to obtain a mirror-finish surface. The HAp ceramics were then immersed into c-SBF at a constant temperature of 310 K. The SBF was not renewed during this study. After 12, 24, 48 and 72 hours of immersion, the top surface and cross-section of the HAp ceramics were observed using SEM (VE-7800, Keyence Corp.) to evaluate the precipitation behavior on the surface.

3. Results and Discussion

3.1. Evaluation of Crystal Orientation and Grain Size Distribution

Figure 1 shows XRD patterns of the three HAp ceramics prepared. For the HAp ceramic fabricated in the absence of a magnetic field (Fig. 1(a)), the XRD pattern had reflections of various crystal faces and the high intensity peaks were very similar to that of HAp (JCPDS #73-0293), which indicates that randomly oriented HAp was successfully fabricated. However, the XRD patterns of the two HAp ceramics slip cast under magnetic fields were significantly different to that of the randomly oriented HAp ceramic. HAp ceramic fabricated under a vertical magnetic field (Fig. 1(b)) displayed an intense (300) reflection corresponding to the a-face, and c-face reflections were of very low intensity compared to that of the randomly oriented one. On the other hand, HAp ceramic fabricated under a horizontal magnetic field with mechanical mold rotation (Fig. 1(c)) displayed strong (002) and (004) reflections, which correspond to c-faces, and reflections of other faces were significantly suppressed.

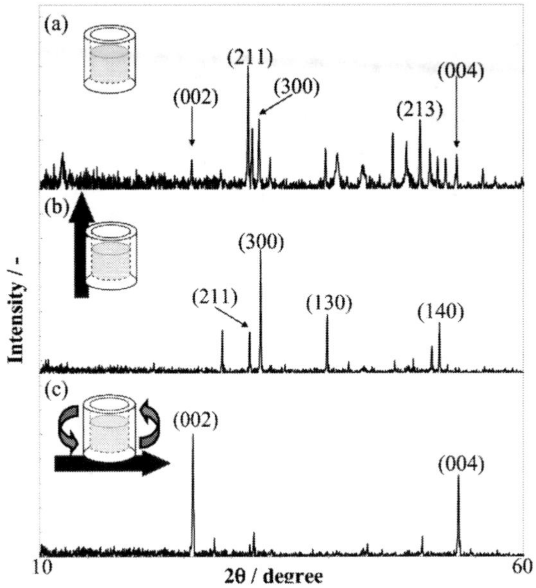

Figure 1. XRD patterns of the three fabricated HAp ceramics
(a) Slip cast without magnetic field
(b) Slip cast with vertical magnetic field
(c) Slip cast with horizontal magnetic field and mold rotation

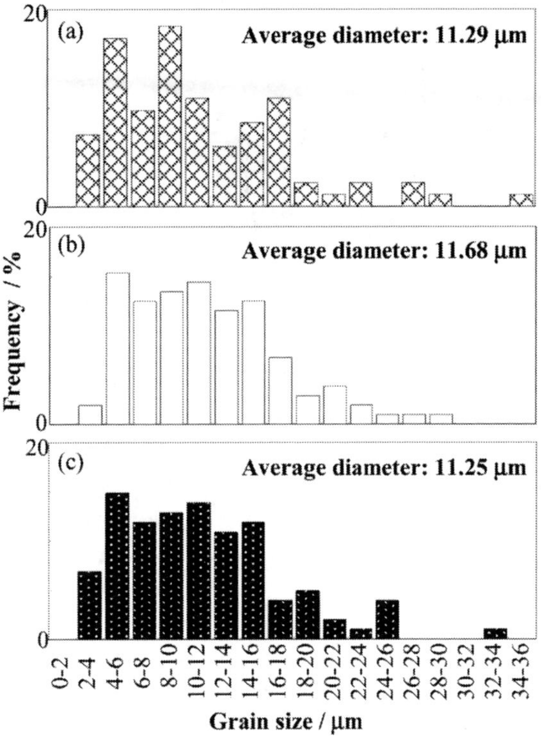

Figure 2. Grain size distribution of the three HAp ceramics
(a) Randomly oriented HAp ceramic
(b) A-face oriented HAp ceramic
(c) C-face oriented HAp ceramic

These results indicate that the two HAp ceramics fabricated with a magnetic field had high crystal orientation to the a-face or c-face.

Figure 2 shows the grain size distribution of each HAp ceramic prepared in this study. More than 90% of the total grains had a diameter between 2 and 20 μm, with an average diameter around 11.5 μm for all of the three HAp ceramics, indicating that no significant differences were observed with respect to grain size and distribution.

3.2. Dependence of Precipitation Behavior on the Crystal Orientation of HAp

Figures 3 (a)-(c) show SEM images of the top surface of the three HAp ceramics immersed for 12 hours in SBF. No obvious precipitation was observed on the a-face oriented HAp ceramic; however, precipitation was evident on the randomly oriented and c-face oriented HAp ceramics. Moreover, the boundaries of the precipitated areas were composed of not curves, but lines on the top surface of the randomly oriented HAp. The precipitation behavior might depend on individual grains, because the precipitated areas were in polygonal shapes that resembled the shape and size of individual grains observed on the etched surface of the randomly oriented HAp ceramic, as shown in Fig. 4.

The variation in the thickness of the top surface precipitate that formed on the a-face and c-face oriented HAp ceramics immersed in SBF was evaluated over time for 12, 24, 48 and 72 hours. The average thickness of the precipitate was calculated from sixteen randomly selected positions measured from SEM cross sectional images. The calculated values are shown in Fig. 5. The precipitate became much thicker on the c-face oriented HAp ceramic compared to the a-face oriented one, which may indicate that the c-face has a greater ability to form a precipitate than the a-face. The thickness of the precipitate on the c-face oriented HAp ceramic seemed to become saturated at 48 hours of SBF immersion. Furthermore, precipitate formation occurred within a shorter time on the c-face oriented HAp than on the a-face oriented HAp. These results suggest that the c-face will bond to living bone in a shorter period than the a-face.

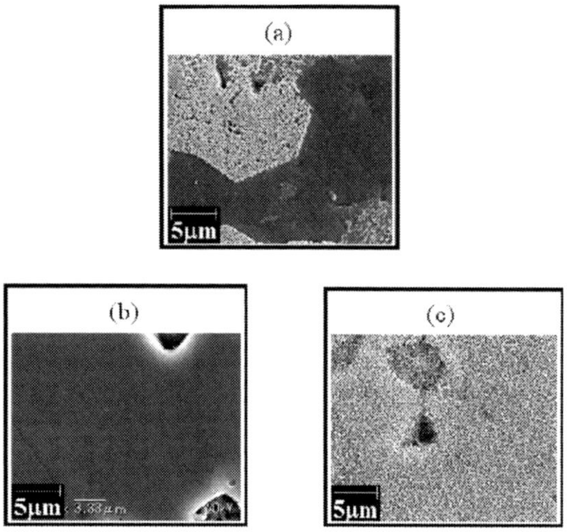

Figure 3. SEM images of the top surface of each crystal oriented HAp ceramic after 12 hours of SBF immersion
(a) Randomly oriented HAp ceramic
(b) A-face oriented HAp ceramic
(c) C-face oriented HAp ceramic

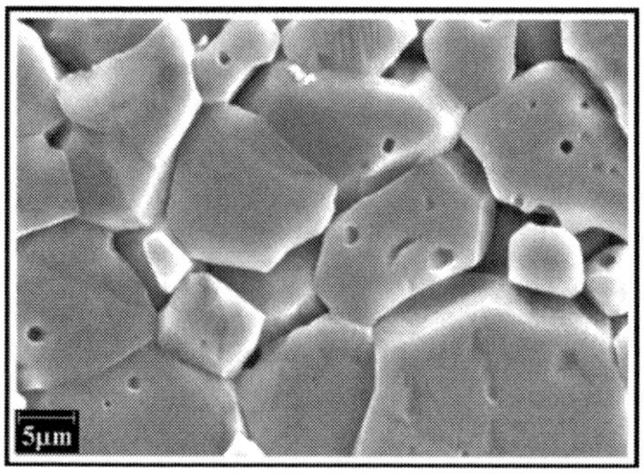

Figure 4. SEM images of the top surface of etched randomly oriented HAp ceramic without immersion into SBF

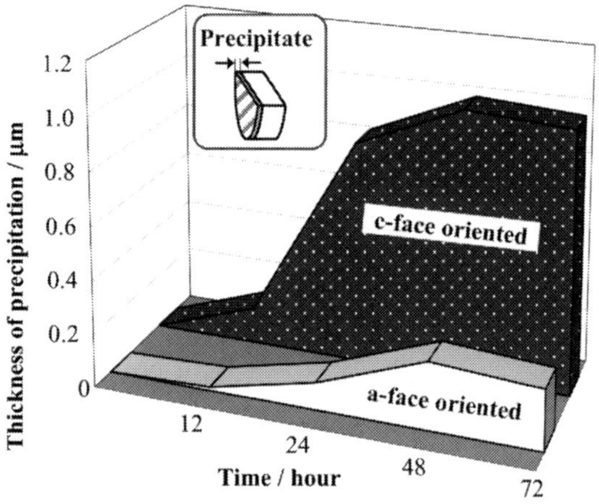

Figure 5. Time dependency of the thickness of precipitate deposited on a-face oriented and c-face oriented HAp ceramics

The difference in the precipitation time and thickness on a-face oriented and c-face oriented HAp may be due to the difference in the surface charge on the HAp faces [13-15]. Taking in account the results of Yamashita et al.[32], Tanahashi et al.[33] and Kim et al.[34], we may conclude that the faster and thicker precipitate formation on the c-face oriented HAp ceramic is due to the negatively charged surface, which strongly attracts calcium ions present in the SBF and accelerates precipitate growth. Slower and thinner precipitate formation on a-face oriented HAp is possibly due to the positively charged surface, which attracts chloride ions and restricts crystal growth, as mentioned in Reference 32. Considering this, the reason why the thickness of precipitate on the c-face oriented HAp exhibited a saturation plateau could be explained as follows. The precipitation rate on the c-face oriented HAp was initially fast, so that the concentration of calcium ions in the SBF must have decreased, leading to a reduction of the precipitation rate. Onuma et al. [35-37] reported that the growth mechanism was dependent on the a-face and c-face orientation of HAp. According to his study, the growth mechanism on the a-face was a combination of step flow and two-dimensional nucleation, while the growth on the c-face followed a multiple two-dimensional nucleation mechanism. These different growth mechanisms might also contribute to the significant difference in the thickness of the precipitate produced during SBF immersion.

4. Conclusion

HAp ceramics with controlled crystal orientation were prepared by a slip casting method under a magnetic field. According to the results of XRD and SEM observations, the fabricated HAp ceramics had surfaces mainly consisting of a-face or otherwise c-face HAp, with no significant difference in grain size and grain size distribution. An experimental investigation of the bioactivity dependency on the two typical crystal faces of HAp was conducted in a simulated body environment provided by SBF.

It was concluded that the thickness of the precipitate was increased and formed earlier on the HAp c-face compared to that on the a-face.

Acknowledgement
This work was partially supported by JSPS Asian Core Program "Construction of the World Center on Electromagnetic Processing of Materials"

References

[1] Jarcho M 1981 *Clin. Orthop.* **157** 259
[2] Ioku K 1996 *J. Soc. Inorg. Mater. Japan* **3** 412
[3] Oonishi H 1991 *Biomaterials* **12** 171-178
[4] Aoki H 1994 *Medical Applications of Hydroxyapatite* (Ishiyaku EuroAmerica Inc., Tokyo, St. Louis) p1-12
[5] Matsumoto T, Okazaki M, Inoue M, Hamada Y, Taira M and Takahashi J 2002 *Biomaterials* **23** 2241-2247
[6] Hlady V and Füredy-Milhofer H 1979 *J. Colloid Interface Sci.* **69** 460
[7] Kandori K, Shimizu T, Yasukawa A and Ishikawa T 1995 *Biointerfaces* **5** 81-87
[8] Kawachi G, Sasaki S, Nakahara K, Ishida E H and Ioku K 2006 *Key Eng. Mater.* **309-311** 935-938
[9] Aoki H and Kato K 1975 *Ceramics Japan* **10** 469-478
[10] Hjerten S 1956 *Biochem. Biophys. Acta* **31** 216
[11] Kawasaki T, Ikeda I, Takahashi A and Kobuki Y 1986 *Eur. J. Biochem.* **155** 249
[12] Kawasaki T, Ikeda I, Takahashi A and Monma H 1986 *Eur. J. Biochem.* **157** 29
[13] Kawasaki T, Takahashi S and Ikeda K 1985 *Eur. J. Biochem.* **152** 361-371
[14] Kawasaki T 1978 *J. Chromatogr.* **151** 95-112
[15] Kawasaki T 1978 *J. Chromatogr.* **157** 7-42
[16] Morikawa H, Sassa K and Asai S 1998 *Mater. Trans.* **39** 814-818
[17] Asai S, Sassa K and Tahashi M 2001 *Sci. Thecnol. Adv. Mater.* **4** 455-460
[18] Kimura T 2003 *Polymer Journal* **35** 823-843
[19] Suzuki T, Sakka Y and Kitazawa K 2001 *J. Ceram. Soc. Japan* **109** 886-890
[20] Tanaka S, Makiya A, Kato Z, Uchida N, Kimura T and Uematsu K 2006 *J. Mater. Res.* **21** 703-707
[21] Inoue K, Sassa K, Yokogawa Y, Sakka Y, Okido M and Asai S 2003 *Mater. Trans.* **44** 1133-1137
[22] Sakka Y, Takahashi K, Mastuda N and Suzuki T S 2007 *Mater. Trans.* **48** 2861-2866
[23] Sakka Y, Takahashi K, Suzuki T S, Ito S and Matsuda N 2008 *Mater. Sci. and Eng.* A **475** 27-33
[24] Akiyama J, Hashimoto M, Takadama H, Nagata F, Yokogawa Y, Sassa K, Iwai K and Asai S 2005 *Mater. Trans.***46** 203-206
[25] Akiyama J, Hashimoto M, Takadama H, Nagata F, Yokogawa Y, Sassa K, Iwai K and Asai S 2005 *Mater. Trans.* **46** 2514-2517
[26] Nakano T, Kaibara K, Tabata Y, Nagata N, Enomoto S, Marakawa E and Umakoshi Y 2002 *Bone* **31** 479-487
[27] Ohtsuki C, Kushitani H, Kokubo T, Kotani S and Yamamuro T 1990 *J. Biomed. Mater. Res.* **24** 721- 734
[28] Cho S -B, Nakanishi K, Kokubo T, Soga N, Ohtsuki C, Nakamura T, Kitsugi T and Yamamuro T 1995 *J. Am. Ceram. Soc.* **78** 1769-1774
[29] Ohtsuki C, Aoki Y, Kokubo T, Bando Y, Neo M and Nakamura T 1995 *J. Ceram. Soc. Japan* **103** 449-454
[30] Kim H -M, Miyaji F, Kokubo T, Nishiguchi S and Nakamura T 1999 *J. Biomed. Mater. Res.* **45** 100-107

[31] Kokubo T and Takadama H 2006 *Biomaterials* **27** 2907-2915
[32] Yamashita K, Oikawa N and Umegaki T 1996 *Chem. Mater.* **8** 2697-2700
[33] Tanahashi M and Matsuda T 1997 *J. Biomed. Mater. Res.* **34** 305-315
[34] Kim H M, Himeno T, Kokubo T and Nakamura T 2005 *Biomaterials* **26** 4366-4373
[35] Onuma K, Ito A, Tateishi T and Kameyama T 1995 *J. Cryst.Growth* **154** 118-125
[36] Kanzaki N, Onuma K, Ito A, Teraoka K, Tateishi T and Tsutsumi S 1998 *J. Phys. Chem.* B **102** 6471-6476
[37] Onuma K, Kanzaki N, Ito A and Tateishi T 1998 *J. Phys. Chem.* B **102** 7833-7838

Alignment of carbon nanotubes by magnetic fields and aqueous dispersion

B K Jang[1], Y Sakka[1], S K Woo[2]

[1]Nano Ceramics Center,National Institute for Materials Science
1-2-1 Sengen, Tsukuba, Ibaraki 305-0047, Japan
[2]Convergence Materials Research Center, Korea Institute of Energy Research
102 Gajeong-ro, Yuseong-gu, Daejeon, 305-343, Korea

E-mail: JANG.Byungkoog@nims.go.jp

Abstract. Homogeneous dispersion of multi-walled carbon nanotubes (CNTs) in aqueous solution was achieved using several dispersants. The most efficacious dispersant was Polyethyleneimine (PEI). Stable CNT dispersions are found to have higher zeta potentials compared to poorly dispersed suspensions. Application of a strong magnetic field of 12 T to CNT to align in the direction of the magnetic field. Well-aligned CNTs according to direction of magnetic field were obtained.

1. Introduction

Carbon nanotubes (CNTs) have attracted great interest because of their unique structural, electronic, physical and thermal properties, such as high electrical conductivity, thermal conductivity and elastic modulus. Consequently, CNTs have been widely studied as additives or reinforcing materials in advanced composites, including epoxy, metal, ceramic and nano-composites [1-5].

It is well known that CNTs with high aspect ratios easily form aggregates or bundles owing to very strong van der Waals interactions between them [6]. It is therefore difficult to prepare stable dispersions of CNTs in aqueous or organic solution [7, 8]. However, the alignment of CNTs and formation of stable dispersions is necessary to optimize the properties of technologically useful materials, e.g., reinforced composites, electronic components and optical materials, so considerable effort has been expended in finding ways of forming stable dispersions of well-aligned CNTs.

To date, several methods have been reported that produce stable dispersions of CNTs in aqueous or organic solution, such as use of dispersing agents [9-11], colloidal surfactants [12] and chemical treatments [13, 14]. The use of an appropriate dispersing agent is particularly attractive as it minimizes the chemical damage or modification of CNT surfaces compared with the other methods.

In addition, methods for aligning CNTs include use of templates [15,16], self-standing geometries [17], and electric fields [18]. However, these methods have not been investigated in detail. There are still some challenges that need to be overcome to enable fully controlled alignment of CNTs. It has been reported that high magnetic fields are useful for producing a controlled texture in anisotropic ceramics such as Al_2O_3, AlN and hydroxyapatite [19, 20]. Therefore, we focus on the use of high magnetic field for alignment of CNTs [21]. The purpose of this work is to investigate the dispersion of multi-walled CNTs and alignment of CNTs using high magnetic field in aqueous solution.

Journal of Physics: Conference Series **156** (2009) 012005 doi:10.1088/1742-6596/156/1/012005

2. Experimental

2.1. Preparation of CNT suspensions

Commercial multi-walled CNT powders (IlJin Co. Korea) that were synthesized by catalytic chemical vapor deposition (CVD) were used in this study. The properties of the pristine CNT powders are listed in Table 1; the powders were found to be 95 wt% pure. Polyethyleneimine, ammonium hydroxide (NH_4OH) and sodium hydroxide (NaOH) were used as dispersing agents in aqueous solution in this study. 1~5 mg of CNTs and 50~500 μl of each dispersant were added to 30 ml of distilled water. Ultrasonication was used to help to break up the agglomerates by providing sufficient energy to overcome the attractive van der Waals forces between nanotubes [22,23]. Ultrasound generator was input into each suspension vials and the mixture was simultaneously stirred using a magnetic bar under ultrasonication for 10~30 mins to disperse and to break the bundle or agglomerates of the CNTs.

2.2. Characterization

Particle size distributions of the CNTs were measured using a laser particle analyzer (Otsuka Electronics, LSPZ-100). Phase analysis of the CNTs was also performed by X-ray diffraction (Jeol, JDX-3500) using CuK_α radiation. The zeta potential of each dilute suspension was measured with a zeta potential analyzer (Otsuka Electronics, LSPZ-100 series). Each sample was ultrasonicated for 10 mins prior to measurement. The ionic strength was maintained using a 10 mM NaCl solution. The zeta potentials (ζ) were calculated from [24]

$$\zeta = VS/\Delta p \cdot \eta/\varepsilon\varepsilon_0 \cdot L/A \cdot 1/R \qquad (1)$$

where VS is the streaming potential, Δp the hydrodynamic pressure difference across the sample, η is the viscosity and ε the permittivity of the liquid, ε_0 the permittivity of free space, L and A are the length and cross sectional area of the sample, respectively, and R is the electrical resistance across it.
After ultrasonication, a drop of each suspension was carefully placed on a silicon wafer using a micropipette for microstructural analysis. The suspension was dried at room temperature for 24 hrs. The morphology and conformation of the CNTs was characterized by digital microscopy (Keyence, VHX-600) and scanning electron microscopy (SEM) (Jeol, JSM-6500).

2.3. Application of a high magnetic field

A strong magnetic field of 12 Tesla was applied to the suspensions to seek to align the CNTs. The droplets of the CNT suspensions on silicon wafers were carefully placed within the magnetic generator equipment at the (horizontal) center of the magnetic field for 1~10 hrs as illustrated in Fig.1.

Table 1. Properties of CNT powders.

Property	Value
Diameter (nm)	10~15
Length (μm)	10~20
Purity (%)	>95
Surface area (m^2/g)	200
Bulk density (g/cc)	0.1

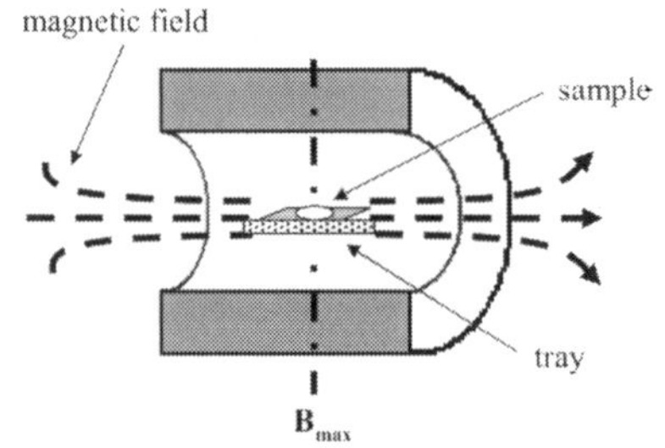

Figure 1. Schematic diagram of setting of sample in magnetic field generator.

3. Results and discussion

3.1. Dispersion of CNTs

Fig. 2 shows typical micrographs of raw powders of CNTs. The CNTs appear to be solid particles (Fig. 2(a)). This indicates that the nanotubes have agglomerated/flocculated to form particle-like bundles because of the relatively strong van der Waals interactions between CNTs. The surfaces of the CNT bundles have the appearance of soft cotton or fine hair (Fig. 2(b)). The laser particle analyzer gave particle sizes of about 272 nm.

A typical XRD pattern of the CNT powders is shown in Fig. 3. This confirms that the observed phase is pure carbon, with the main peak at $2\theta = 25.6$ degrees [25].

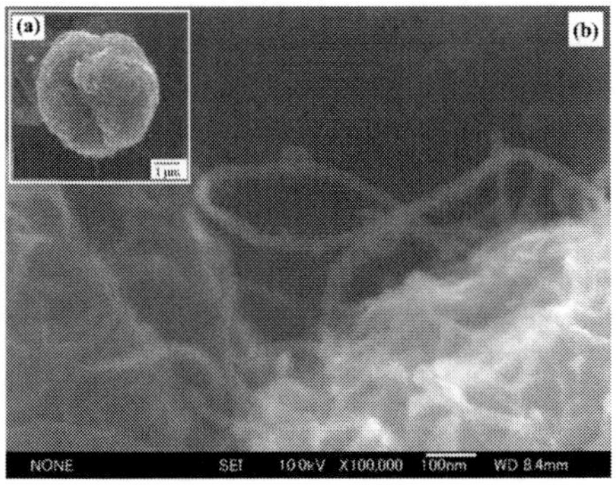

Figure 2. Microstructure of CNT powders; (a) and (b) are low magnification and high magnification of surface of (a), respectively.

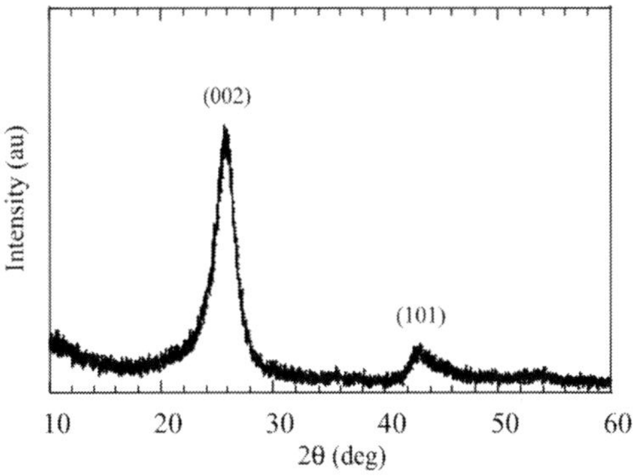

Figure 3. XRD pattern of CNT powders.

Fig. 4 shows photographs of vials containing 0.1 mg CNT dispersions after mixing by ultrasonic agitation. Samples A, B and C are inhomogeneous, with considerable sedimentation towards the bottom of the vials, resulting in the rest of the dispersion being more-or-less completely transparent. CNTs of their suspensions were insoluble even after ultrasonication. This flocculation is a result of the significant van der Waals interaction between the sidewalls of the CNTs.

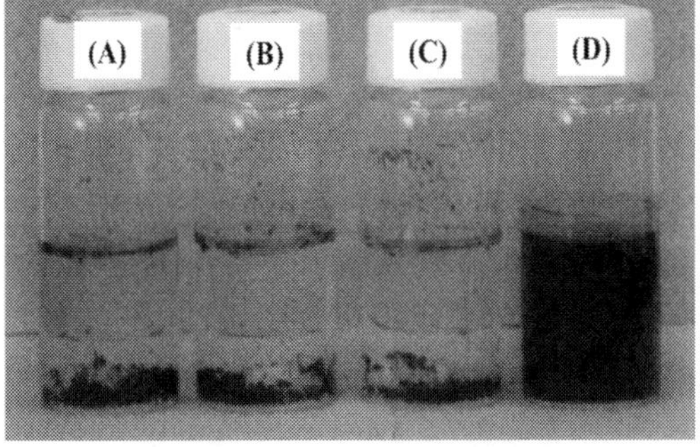

Figure 4. Influence of dispersant on CNT dispersions in aqueous solution:
(A) no dispersant; (B) 1N-NH_4OH; (C) 5N-NaOH, and (D) PEI as dispersants.

However, sample D corresponded to a homogeneous, well-dispersed CNT suspension black in appearance and with no observable precipitation. The CNTs remained in the dispersed state for over one year. This result demonstrates that PEI is an effective dispersant for CNTs. The strong electrostatic repulsion introduced between the nanotubes by the organic molecules successfully separate the nanotubes for long periods of time.

To investigate the dispersion behavior of CNTs in aqueous solutions, zeta potentials were measured. The zeta potential and pH values of each sample are listed in Table 2. The first three samples have negative zeta potentials varying around -15 mV, whereas sample D has a zeta potential of +20 mV. It is known that pristine CNT suspensions without dispersants have a minus charge at pH of 8.5. In addition, it is evident that the CNTs are negatively charged in aqueous solution. The addition of dispersants contribute to increase of the magnitude of zeta potential and to make the alkaline solution by increasing the pH values from 8.5 to 11.8.

Table 2. Zeta potential and pH of CNT dispersed samples in aqueous solution.

Sample	Zeta potential (mV)	pH
Sample A	-13.7	8.5
Sample B	-14.6	9.6
Sample C	-18.5	11.8
Sample D	20.3	9.2

The effect of the PEI for stable suspension is follows. The PEI can be wet to surfaces of CNTs and create the repulsive potential. Electrostatic repulsion between the relatively charged CNT surfaces plays useful role in the stabilization of the CNT clusters by van der Waals attraction in aqueous solution. It was reported that the PEI in ceramic colloidal dispersion attributed to stabilize the ceramic aqueous suspension such as Al_2O_3, SiO_2 by the electrostatic repulsion [26].

Fig. 5 shows typical SEM micrographs of dispersed CNTs. The CNT morphology of Fig. 5 (a) consists of huge agglomerates of CNTs because of the mutual attraction between nanotubes. In contrast, the CNTs in Fig. 5 (b) are well dispersed and individually distinguishable. Fig. 5 (b) does not show any huge agglomerates, although there are still small agglomerates visible (indicated by arrows). The well-dispersed CNTs are randomly distributed like individual strands of hair. This result clearly demonstrates that PEI is a suitable dispersant for separating CNTs in suspension.

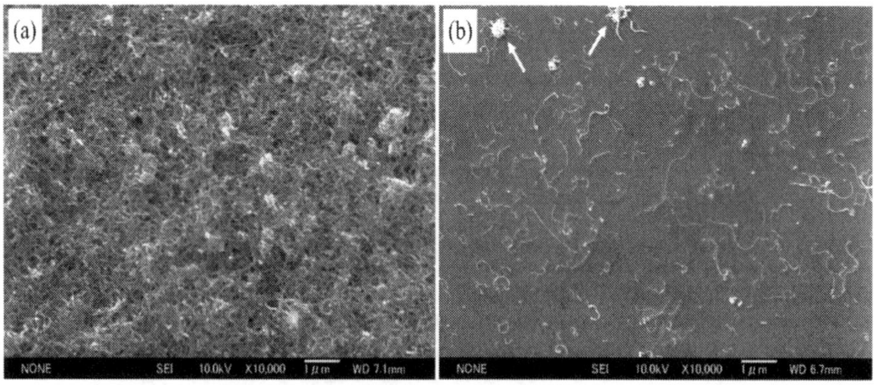

Figure 5. Microstructure of dispersed CNTs by SEM; (a) sample A and (b) sample D.

3.2. Alignment of CNTs

Fig. 6 shows SEM micrograph at high magnification after applying a magnetic field to obtain the aligned CNTs. In Fig.6, small curved CNTs show a tendency to align. However, the larger curved CNTs are not as well aligned. It might reasonably be expected that straight CNTs will align more easily than curved CNTs.

The aligning mechanism of ceramic particles by a strong magnetic field can be explained as follows [19,20]. Two conditions are necessary for the particles to align, namely, good dispersion and an anisotropic structure of the particles. Particles with an anisotropic magnetic susceptibility can rotate to an angle that minimizes the system energy when a magnetic field is applied. The energy induced by anisotropic magnetic susceptibility is expressed as follows [20]:

$$\Delta E = - \Delta\chi V B^2/2\mu_0 \quad (2)$$

where $\Delta\chi = \chi_{a,b}-\chi_c$ is the anisotropy of the magnetic susceptibility, μ_0 is the permeability in vacuum, B is the applied magnetic field and V is the volume of each particle. Alignment of anisotropic particles occurs when ΔE is higher than the thermal energy, $k_B T$.

The present result suggests that application of a magnetic field is very effective for aligning CNT particles. Further development of this method is expected to lead to the development of nanocomposites with superior structural and functional properties.

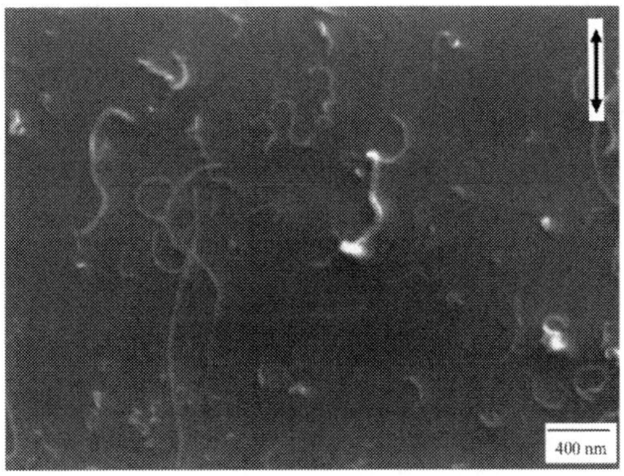

Figure 6. Micrographs of CNT alignment under a magnetic field of sample D. (Arrows indicate the direction of the magnetic field).

4. Conclusions

Alignment of multi-walled CNTs using a strong magnetic field on a dispersion of CNTs in aqueous solution was successfully achieved. Polyethyleneimine was shown to be an effective dispersant for preparing the dispersion. Well-dispersed suspensions of CNTs have higher zeta potentials than poorly-dispersed suspensions. The aligned CNTs according to direction of magnetic field were successfully obtained. The degree of alignment of the CNTs depended on the shape of the CNTs.

Acknowledgments

This work was supported by a seed grant from National Institute for Materials Science (NIMS) and by grant of contract research with Korea Institute of Energy Research (KIER).

References

[1] Zhou Y, Pervin F, Lewis L, Jeelani S 2008 *Mater. Sci. Eng.* A **475** 157
[2] Li K , Saigal S 2007 *Mater. Sci. Eng. A* **457** 44
[3] Chen W X, Tu J P, Wang L Y, Gan H Y, Xu Z D, Zhang X B 2003 *Carbon* **41** 215
[4] Xia Z H, Lou J, Curtin W A 2008 *Scripta Mater.* **58** 223
[5] Robel I, Bunker B A, Kamat P V 2005 *Adv. Mater.* **17** 2458
[6] Bonard J M, Stora T, Salvetat J P, Maier F, Stockli T, Duschi C, Forró L, de Heer W A, Châtelain A 1997 *Adv. Mater.* **9** 827
[7] Zhu J, Yudasaka M, Zhang M, Iijima S 2004 *J. Phys. Chem. B* 108 11317
[8] Lee G W, Kumar S J 2005 *J. Phys. Chem. B* **109** 17128
[9] Jang B K, Suzuki T S, Uchikoshi T, Tsuda K, Sakka Y 2008 *32th International Cocoa Beach Conference & Exposition on Advanced Ceramics and Composites*, APMT-01
[10] Yu J, Grossiord N, Koning C E, Loos J 2007 *Carbon* **45** 618
[11] Hungerbühler H, Guldi D M, Asmus K D 1993 *J. Am. Chem. Soc.* **115** 3386
[12] Sun J, Iwasa M, Gao L, Zhang Q 2004 *Carbon* **42** 885
[13] Lu C, Chiu H 2008 *Chem. En. J.* **139** 462
[14] Saito T, Matsushige K, Tanaka K 2002 *Physica B* **323** 280
[15] Ji L, Wang Z, Li Z, Liang J 2008 *Mater. Lett.* **62** 1979
[16] Xu J, Zhang X, Chen F, Li T, Li Y, Tao X, Wang Y, Wu X 2005 Appl. *Surf. Sci.* **239** 320
[17] Dasgupta K, Kar S, Venugopalan R, Bindal R C, Prabhakar S, Tewari P K, Bhattacharya S, Gupta S K, Sathiyamoorthy D 2008 *Mater. Lett.* **62** 1989
[18] Ma C, Zhang W, Zhu Y, Ji L, Zhang R, Koratkar N, Liang J 2008 *Carbon* **46** 706
[19] Suzuki T S, Uchikoshi T, Sakka Y 2007 *Mater. Transaction* **48** 2883
[20] Sakka Y, Suzuki T S 2005 *J. Ceram. Soc. Jpn.* **113** 26
[21] Fujiwara A, Ootoshi H, Suematsu H, Yumura M, Uchida K 1998 *Proc. of Super Carbon (IUMRS-ICA-97)*, Tokyo 73
[22] Sano M, Kamino A, Okamura J, Shinkai S 2001 *Science* **293** 1299
[23] Bonard J M, Stora T, Salvetat J P, Maier F, Stöckli T, Duschl C, Forró L, Heer W A, Châtelain A 1997 *Adv. Mater.* **9** 827
[24] Christoforou C C, Clark G B W, Anderson J L 1985 *J. Coll. Interface Sci.* **106** 1
[25] Jiang Z, Song R, Bi W, Lu J, Tang T 2007 *Carbon* **45** 449
[26] Baklout S, Pagnoux C, Chartier T, Baumard J F 1997 *J. Eur. Ceram. Soc.* **17** 1387

Journal of Physics: Conference Series **156** (2009) 012006 doi:10.1088/1742-6596/156/1/012006

Electro-Migration Effect of Titanium Structure under High Magnetic Field

Y Sakaida[1], K Iwai[1] and S Asai[2]

[1]Department of Materials, Physics and Energy Engineering, Nagoya University 464-8603 Furo-cho, Chikusa-ku, Nagoya, Japan

[2]Innovation Plaza Tokai Japan Science and Technology Agency・Director

E-mail: sakaida.yasuhiko@d.mbox.nagoya-u.ac.jp

Abstract. It is well known topic that electro-migration accompanies the transport of atoms, ions and holes in a material. Then we can say that both electro-migration and strong magnetic field have potential to control crystal alignment. In this study, an electric current and/or a static magnetic field were applied to a titanium sample parallel to each other to investigate the effects of the electric current and the static magnetic field on crystal alignment in the titanium. Using XRD analysis, it was understood that titanium crystals having a c-axis perpendicular to the electric current direction increased under the imposition of only the electric current and under the simultaneous imposition of the electric current and the magnetic field. On the other hand, titanium crystals having a c-axis parallel to the magnetic field direction increased under the imposition of only the magnetic field. The aligned result can be explained from the viewpoint of anisotropy of the Joule loss and the magnetization energy.

1. Introduction

The imposition of an electric current on electrically conductive materials such as metals and alloys induces electro-migration and thermal strain in the materials, leading to deterioration and loss of continuity in electric circuits. These phenomena have been investigated mainly from the viewpoint of the reliability of integrated circuits so far, because the electric current density in LSI (large-scale integrated circuit) and ULSI (ultra-large scale integrated circuit) has drastically increased. For example, simulations and computer models [1], temperature dependence of electro-migration [2], and transformation behavior of solder alloys [3], have been performed. In these studies, copper, aluminum, and solder alloys have been treated as the test materials usually used in integrated circuits.

Conversely, an electric current has the potential to control the structure of a material because the transport phenomena of atoms, ions, and holes in a material can be controlled through electro-migration. For example, electro-migration has been applied to refining of an electrically conductive material [4,5].

On the other hands, a magnetic field is well known as a powerful tool for controlling crystal alignment of crystallographically anisotropic materials [6-8]. In addition, a strong magnetic field affects concentration of an additional element at grain boundary in a solid alloy through an annealing process [9,10]. This suggests that a strong magnetic field might affect diffusion behavior and these phenomena might introduce anisotropic growth of a solid metal or alloy. However, the effect of the imposition of an electric current and/or a magnetic field on crystal alignment had not been discussed until now.

In this study, an electric current and/or a static magnetic field were applied to a titanium sample simultaneously parallel to each other to investigate the effect of the electric current and the static magnetic field on crystal alignment in the sample.

2. Experimental Procedure and Result

Titanium is chosen as a sample because it shows anisotropic nature in electric conductivity and magnetic susceptibility caused by its hexagonal crystallographic structure. In this study, the electric current was applied to a solid titanium sheet under the imposition of the strong magnetic field generated by a super conducting magnet. The magnetic field direction was parallel to the electric current direction as shown in Fig.1. The obtained structure was analyzed by XRD for the evaluation of the crystal orientation. To reduce the temperature change during the experiment, the sample was cooled by ion-exchanged water. In this experiment, two titanium sheets were used. The sample1, 2 and 3 were cut out from one titanium sheet, and the sample1' and 4 were cut out from another one.

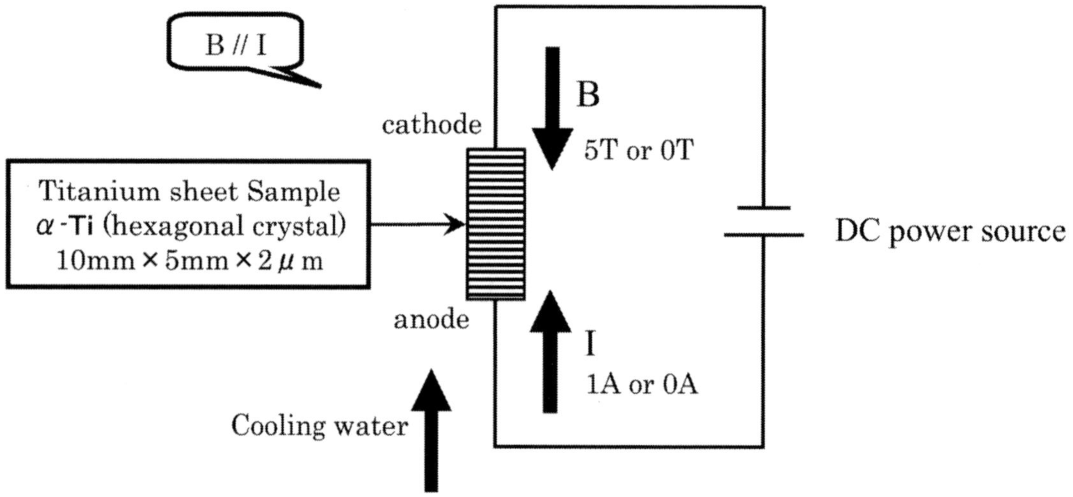

Figure 1. Experimental apparatus

In the Debye Rings of every sample, three peaks of (100), (002) and (101) planes are mainly observed as shown in Fig.2. Cross-sectional profile at ϕ =0 degree is called the horizontal profile, while that at ϕ =90 degrees is called the vertical profile in this paper. The magnetic field direction and electric current direction are same with the direction of vertical profile in the Debye Rings shown in Fig.2.

Because the intensity of each ring is not uniform in azimuthal direction, titanium crystals in each sample are aligned to the specific direction.

As an index of crystal alignment, we introduced the peak intensity ratio of (002) plane peak intensity in vertical profile to that in horizontal profile, and evaluated ratio is shown in Fig.3. The intensity was evaluated by integrating the (002) profile with respect to the azimuthal direction.

Because sample1 and 1' were cut out from different titanium sheet each other, the peak intensity ratio of sample 1 and 1' in Fig. 3 were not the same value each other though both the samples were bought from the same company(THE NILACO CORPORATION: Titanium Foil, TI-453098). This might be because the rolling conditions for formation of the titanium samples before we bought were different. The intensity ratio decreased in the case of the imposition of only the electric current. This means that titanium crystals having a c-axis parallel to the electric current relatively decreased in comparison with titanium crystals having a c-axis perpendicular to the electric current. In the case of the imposition of both the electric current and the magnetic field, the experimental result was similar to that of the imposition of only the electric current though decrease of the intensity ratio was small in comparison with the result under the imposition of only the electric current. On the other hand, we can recognize that the intensity ratio slightly increased after the imposition of only the magnetic field as the intensity ratio of sample 4 increased compared to that of the sample 1'. Therefore, the imposition of the magnetic field might have a function to slightly increase the intensity ratio while the imposition of the electric current decreased the intensity ratio in this experimental condition.

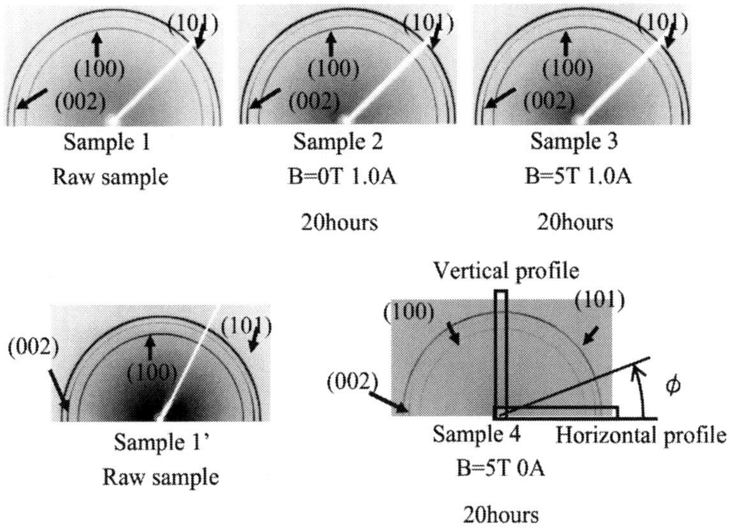

Figure 2. Debye rings of each sample and definition of angle ϕ

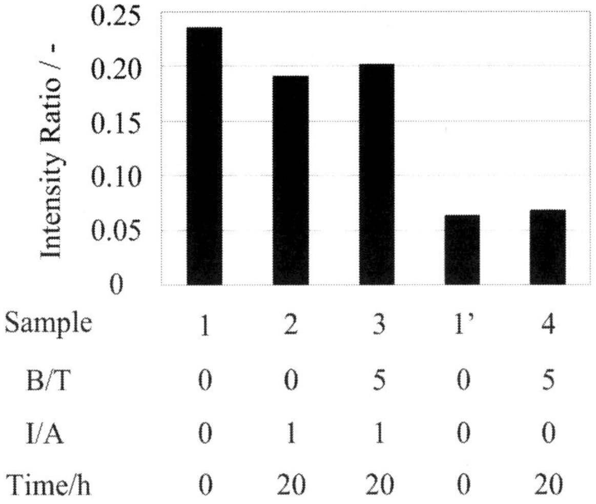

Figure 3. Peak intensity ratio of (002) in vertical profile / (002) in horizontal profile

3. Discussion

Because the magnetic susceptibility of titanium in the c-axis was larger than that in the a-axis when T=300K(Table.1), the adsorption of migrating titanium ions or atoms on a titanium crystal with a c-axis parallel to the magnetic field might occur from the viewpoint of magnetization energy. Then, magnetization energy difference per unit volume U_M can therefore be calculated by the following equation:

$$U_M = -\frac{\mu_0 \Delta\chi}{2}|H|^2 \tag{1}$$

where μ_0 is magnetic susceptibility in a vacuum, $\Delta\chi$ is the magnetic susceptibility difference between magnetically easy axis and magnetically hard axis, and H is the intensity of the external magnetic field.

Table 1. Magnetic susceptibility of titanium

T(K)	//c-axis [-]	⊥c-axis [-]
273	1.99E-04	1.70E-04
300	1.99E-04	1.71E-04
400	2.03E-04	1.77E-04

Table 2. Specific resistance of titanium

T(K)	//c-axis [$10^{-8}\Omega$m]	⊥c-axis [$10^{-8}\Omega$m]
273	47.6	45.35
300	53.8	50.34
400	78.1	71.65

The calculated magnetization energy difference was 300 J/m^3 for titanium at the experimental temperature of 300K. On the other hand, thermal fluctuation (k_BT, k_B: Boltzmann constant, T: absolute temperature) was approximately 10^{-21} J when T = 300K. Therefore, the minimum crystal size in which the magnetization energy difference was larger than the thermal fluctuation was calculated as 3.1×10^{-8}m under these experimental conditions if the interfacial energy was neglected. The increase in the intensity ratio by imposing only the magnetic field can be explained from the viewpoint of the magnetization energy. However, mechanism why the magnetically stable crystals increased and/or grew in the solid metal under the magnetic field is future work while a strong magnetic field might affect diffusion behavior as mentioned in the introduction.

On the other hand, crystal growth whose c-axis is perpendicular to the electric current direction might occur to reduce the Joule loss in the sample, because the specific resistance in the c-axis was larger than that in the a-axis as shown in Table.2. In other words, c-axis of the titanium crystals aligned themselves to become perpendicular to the electric current direction. The Joule loss difference per unit volume, U_E was then calculated using the following equation:

$$U_E = \Delta\rho\left|J^2\right| \tag{2}$$

where $\Delta\rho$ is the specific resistance difference between electrically-easy axis and electrically-hard axis, and J is electric current density.

The calculated Joule loss difference was 3.5×10^8 J/m^3 for titanium at the experimental temperature of 300K. This value was about 10^6 times as large as the calculated magnetization energy difference per unit volume. That is, crystals whose a-axis were parallel to the electric current direction increased even though the magnetic field were simultaneously imposed on the sample from the energy viewpoint.

This agrees with the experimental results obtained under the imposition of only the electric current and the simultaneous imposition of the electric current and the magnetic field. Therefore, this might be the reason why the titanium crystals in the sample aligned to reduce the Joule loss even though the magnetization energy difference was larger than the thermal fluctuation. In these conditions, migration caused by the imposition of the electric current might enhance the growth of the electrically stable crystals. However, further investigation is needed to confirm the mechanism.

4. Conclusion

To investigate the effect of the electric current and the static magnetic field on crystal alignment, an electric current and/or a static magnetic field was applied to titanium samples which have crystallographic anisotropy in electric and magnetic properties. The titanium crystals aligned to reduce the magnetization energy under the imposition of only the magnetic field. On the other hand, the titanium crystals aligned to reduce the Joule loss because the Joule loss was larger than the magnetization energy.

Acknowledgement

This work was partially supported by JSPS Asian Core Program "Construction of the World Centre on Electromagnetic Processing of Materials".

References

[1] Shingubara S, Utsunomiya I and Fujii T 1995 *The IEICE Transactions.* C-II **J78-C-II** 294-304
[2] Christou A 1993 *Electromigration and Electronic Device Degradation* (Wiley-Interscience)
[3] Hung F Y, Wang C J, Lui T S and Chen L H 2005 *Mater. Trans.* **46** 1820-1824
[4] Kuroda K, Horibe Y, Fujisawa T and Yamauchi C 1994 *J.Japan Inst. Metals.***58** 634-641
[5] Sano H, Yoshida T, Kuroda K, Fujisawa T and Yamauchi C 1999 *Shigen-to Sozai* **115** 54-58
[6] Akiyama J, Hashimoto M, Takadama H, Nagata F, Yokogawa Y, Sassa K, Iwai K and Asai S 2005 *Mater. Trans* .**46** 203-206
[7] Akiyama J, Hashimoto M, Takadama H, Nagata F, Yokogawa Y, Sassa K, Iwai K Asai S 2005 *Mater. Trans.* **46** 2514-2517
[8] Akiyama J, Asano H, Iwai K, Asai S. 2007 *J.Japan Inst. Metals.* **71** 108-112
[9] Watanabe T, Tsurekawa S, Zhao X, Zuo L 2006 *Scripta Materialia* **54** 969-975
[10] Wu Y, He C S, Zuo X and Watanabe T 2008 *Acta Materiall. Sin. (Engl. Lett.)* **21** 103-108

Rotational Diffusion Model of 2-Dimensional Magnetic Alignment

M Yamaguchi, S Ozawa, K Ogawa and I Yamamoto

Department of Physics, Yokohama National University, 79-5 Tokiwadai, Hodogaya-ku, Yokohama 240-8501, Japan

E-mail: yamaguch@ynu.ac.jp

Abstract. The dynamic process of the 2-dimensional magnetic alignment was analyzed by the rotational diffusion model which took account of the influence of thermal disturbance. Solving the rotational diffusion equation yielded the distribution function and the order parameter. The influences of diffusion were discussed by comparison with the magnetic alignment without diffusion.

1. Introduction

Applying high magnetic fields causes magnetically weak particles to align, then creates new materials functions.[1] However, usually the magnetic alignment (MA) is disturbed by the Brownian motion of the particles because the magnetic free energy is comparable to the thermal energy in ordinary magnetic fields. Consequently, the MA of such particles show the Boltzmann distribution in direction at equilibrium.[2] This indicates that the influence of Brownian motions must be considered in the dynamic process of the MA also. So far the alignment process was analyzed by the mechanical model which took no account of thermal disturbance. [3,4] Recently, this model is improved by including random torque representing thermal disturbance.[5]

Meanwhile, the influence of Brownian motion on MA is systematically formulated as the rotational diffusion equation (RDE).[6] In this paper, we apply this formulation to the 2-dimensional magnetic alignment (2d-MA) where the rotation of particles is restricted in a plane. The present paper intends to study the essential influences of rotational diffusion on MA, but not to explain a specific experimental result.

2. Formulation

2.1. Rotational diffusion equation

The particle is supposed to have uniaxial anisotropy of magnetic susceptibility $\chi_1=\chi_2\neq\chi_3$ with an ellipsoid of revolution in shape where the principle axes of magnetic susceptibility χ_1, χ_2 and χ_3 coincide with the x_1, x_2 and x_3 axes of ellipsoid, respectively. The anisotropic susceptibility is denoted by $\Delta\chi=\chi_3-\chi_1$. The unit vectors **u**, **v** and **w** are taken along the x_3, x_1 and x_2 axes while **u** is used as the director of a particle. The cylindrical coordinates of **u** are ($r=1$, θ, $z=0$) with respect to the space-fixed axes of X_1, X_2 and X_3 as shown in Fig. 1.

It is presumed that the system consists of N_0 particles which have no interaction with each other. A static magnetic field **B** is applied to the particles along the X_1 axis and their rotational plane is restricted in the X_1X_2 plane. Thus, the potential energy U of the particle depends on the angle θ.

$$U=-\frac{1}{2\mu_0}VB^2\chi_1-\frac{1}{2\mu_0}VB^2\Delta\chi\cos^2\theta=U_0-\beta\cos^2\theta\,, \tag{1}$$

where V is the volume of a particle, U_0 is the θ-independent term and β is the magnetic energy related to the magneto-anisotropy.

$$\beta=\frac{1}{2\mu_0}VB^2\Delta\chi\,. \tag{2}$$

It is well-known that the particle tends to align in parallel or perpendicular to the magnetic field according to the positive or negative $\Delta\chi$.

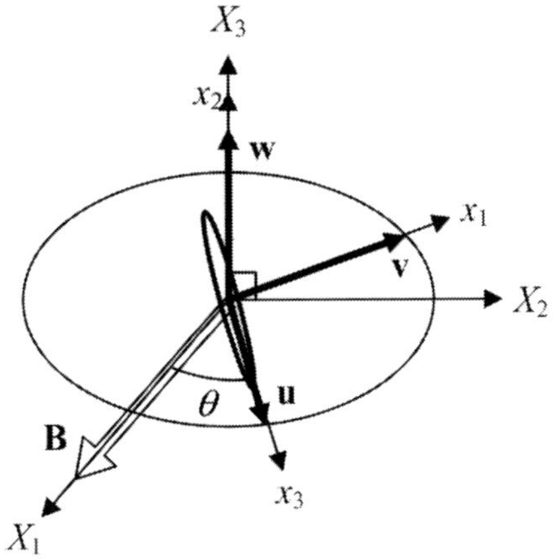

Figure 1. Cylindrical coordinates for the 2-dimensional magnetic alignment of uniaxial particles.

The RDE has been proposed previously.[6]

$$\frac{\partial C}{\partial t}=-\mathbf{R}\cdot\mathbf{J}_R=D_R\mathbf{R}^2C-\mu_R\mathbf{R}\cdot(C\mathbf{N})\,, \tag{3}$$

where C is the concentration of particles, t is the time, **R** is the rotational operator, $\mathbf{J}_R$ is the rotational flux and **N** is the magnetic torque acting upon a particle. D_R and μ_R are the rotational diffusion constant and the rotational mobility, respectively, which are related by the Einstein relation $D_R=k_BT\mu_R$. Eq. (3) can be derived as an analogy to the electric field-induced rotation of polymers.[7] The torque is obtained from the potential energy by $\mathbf{N}=-\mathbf{R}U$. In the present problem, the rotational operator is expressed by $\mathbf{R}=\mathbf{w}\partial/\partial\theta$. We replace the concentration C by the distribution function P by $P=C/N_0$ which is a function of angle θ and time t. Thus, the rotational flux and the RDE for the 2d-MA are written in the cylindrical coordinates as follows.

$$\mathbf{J}_R=\mathbf{w}N_0\left(-D_R\frac{\partial P}{\partial\theta}-\beta\mu_RP\sin 2\theta\right), \tag{4}$$

$$\frac{\partial P}{\partial t}=D_R\frac{\partial^2P}{\partial\theta^2}+\mu_R\beta\frac{\partial}{\partial\theta}(P\sin 2\theta)\,. \tag{5}$$

Alternatively, Eq. (5) can be derived from the viewpoint that point particles move on the arc of a semi-circle with unit radius by the translational diffusion. We take the translational coordinate s along the arc to write the translational diffusion equation.

$$\frac{\partial P}{\partial t} = D_R \frac{\partial^2 P}{\partial s^2} + \mu_R \frac{\partial}{\partial s}(PF). \tag{6}$$

This is the same as Eq. (5) because the force acting on a point particle is $F=\beta\sin 2s$ with $s=\theta$. The distribution function must be normalized by

$$\int_{-\pi/2}^{\pi/2} P d\theta = 1. \tag{7}$$

In addition, it is assumed that the particles are randomly distributed in the initial state at t=0.

$$P(\theta,0) = \frac{1}{\pi}. \tag{8}$$

We introduce the reduced time τ and the field parameter α which represents the ratio of magnetic energy to thermal energy.

$$\tau = D_R t = k_B T \mu_R t , \tag{9}$$

$$\alpha = \frac{\beta \mu_R}{D_R} = \frac{\beta}{k_B T}. \tag{10}$$

Then, the RDE is written by the dimensionless form.

$$\frac{\partial P}{\partial \tau} = \frac{\partial^2 P}{\partial \theta^2} + \alpha \frac{\partial}{\partial \theta}(P \sin 2\theta). \tag{11}$$

Meanwhile, the order parameter $<m>$ is defined for the 2d-MA.

$$<m> = \int_{-\pi/2}^{\pi/2} (2\cos^2\theta - 1) P d\theta. \tag{12}$$

The value of this quantity is 0 for a random distribution, 1 for the perfect parallel alignment along the field and −1 for the perfect perpendicular one.

2.2. Equilibrium state

The equilibrium state is attained when the time tends to infinity, then the time derivative of the distribution function $\partial P/\partial t$ disappears in Eq. (5). Simultaneously, the rotational flux of Eq. (4) becomes zero.

$$\frac{dP}{d\theta} + \alpha P \sin 2\theta = 0. \tag{13}$$

Solving this gives the equilibrium distribution P_{eq}.

$$P_{eq}(\theta) = \frac{\exp(\alpha \cos^2\theta)}{\int_{-\pi/2}^{\pi/2} \exp(\alpha \cos^2\theta) d\theta}. \tag{14}$$

This is a Boltzmann distribution. Substituting Eq. (14) into Eq. (12) gives the order parameter at equilibrium $<m>_{eq}$.

2.3. Diffusion-free system

It is instructive to investigate the hypothetical system without rotational diffusion. This situation replaces Eq. (11) to

$$\frac{\partial P}{\partial \tau} = \alpha \frac{\partial}{\partial \theta}(P \sin 2\theta) \cdot \tag{15}$$

This has the analytical solution with the aid of Eqs. (7) and (8).

$$P = \frac{e^{2\alpha\tau} \sec^2 \theta}{\pi(1 + e^{4\alpha\tau} \tan^2 \theta)}, \quad (\alpha\tau = \beta\mu_{\mathrm{R}} t), \tag{16}$$

Furthermore, this leads to the analytical expression of the order parameter.

$$< m >= \tanh(\alpha\tau) . \tag{17}$$

When the time tends to infinity, $<m>$ becomes 1 for $\alpha>0$ and -1 for $\alpha<0$. In the initial region of time with small τ, this is approximated by

$$< m >= \alpha\tau . \tag{18}$$

It is considered that this diffusion-free system corresponds to the mechanical model without thermal disturbance. [3,4] As a matte of fact, essentially the same relations as Eqs. (16) and (17) have been derived from the mechanical model for the 2-dimensional magnetic alignment by Kimura et al.[8] Comparing Eq. (17) to the corresponding equation Eq. (15A) in ref. 8 gives the relation $\mu_{\mathrm{R}}=(L_1 \eta)^{-1}$ where L_1 is a hydrodynamic coefficient of the particle and η is the viscosity of the ambient medium in the mechanical model.

3. Calculated results and discussion

First, we deal with the MA with rotational diffusion. The distribution function was numerically calculated from Eq. (11) by the difference method for the partial differential equation with Visual C++. The solutions were stably obtained by using the steps of $\Delta\tau$=0.0001 and $\Delta\theta$=1.0 deg with significant figures of three. Figure 2 illustrates the time-variation of the distribution function for a=10 where the curves are drawn only in the positive θ region because of the even function. The figure clearly indicates the time-developing alignment of particles: The distribution function is flat at t=0, then gradually becomes narrower with time and tends to the curve at equilibrium. It must be emphasized that the distribution function at τ=0.4 or more agreed completely with the equilibrium distribution calculated from Eq. (14). This warrants the validity of the computing code used.

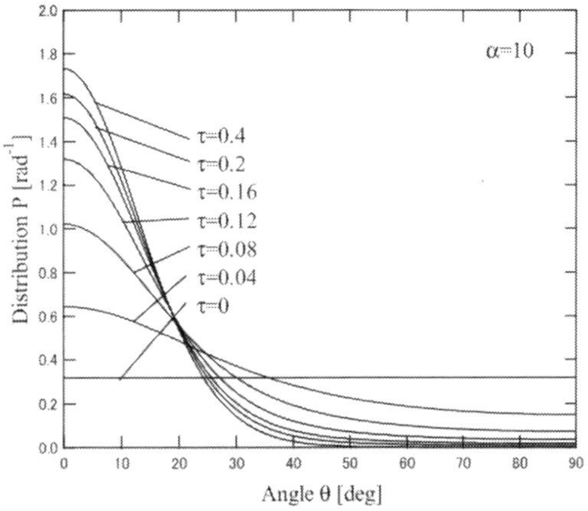

Figure 2. Computed curve of the distribution function P at different times for the 2-dimensional magnetic alignment with rotational diffusion (α=10).

The time-dependence of order parameter $<m>$ is calculated according to Eq. (12) from the computed distribution functions as shown in Fig. 3. Hence, the saturated value of the order parameter agreed with that calculated from Eqs. (12) and (14) for every field parameter. The magnetic alignment attains the equilibrium state faster with larger field parameters α. In addition, it also proceeds faster with larger values of the rotational diffusion constant D_R because the real time is $t=\tau/D_R$. It is recognized that the order parameter is abruptly increased in the region between $\alpha=1$ and 10, meaning that the MA needs the magnetic energy which overcomes the thermal energy.

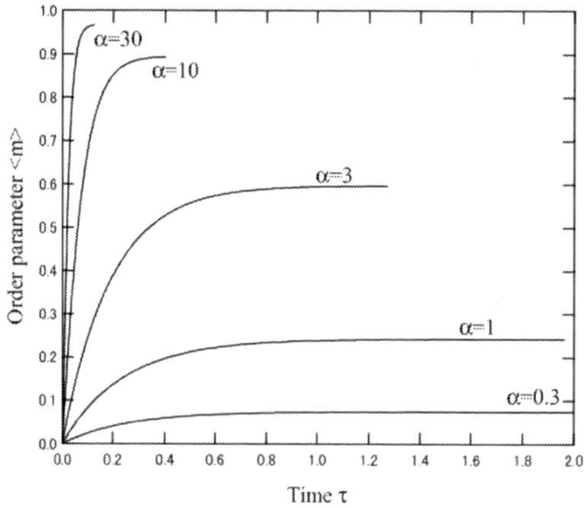

Figure 3. Time-dependence of the order parameter for the 2-dimensional magnetic alignment with rotational diffusion.

Next, as for the MA without rotational diffusion, the distribution function is calculated by the formula of Eq. (16). The distribution function rapidly gets narrower with passing time, which tends to a delta function $P=\delta(\theta)$ at infinite time for every value of α. Then, the order parameter was calculated by the formula of Eq. (17) as shown in Fig. 4. The order parameter necessarily takes unity at infinite time for any value of the field parameter α.

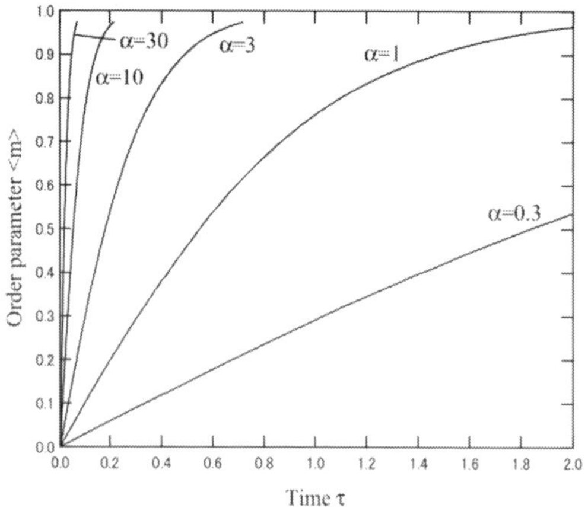

Figure 4. Time-dependence of the order parameter in the 2-dimensional magnetic alignment without rotational diffusion.

In the MA without rotational diffusion, the order parameter varies linearly with time at the initial stage, obeying Eq. (18) as exemplified in Fig. 5. In the MA with rotational diffusion, diffusion is negligible at the initial stage when the particle concentration is nearly homogeneous. Accordingly, the order parameter is approximately linear against time in the MA with rotational diffusion also. As time passes, however, diffusion occurs noticeably because the distribution has a gradient. Then, the magnetic alignment is suppressed by diffusion.

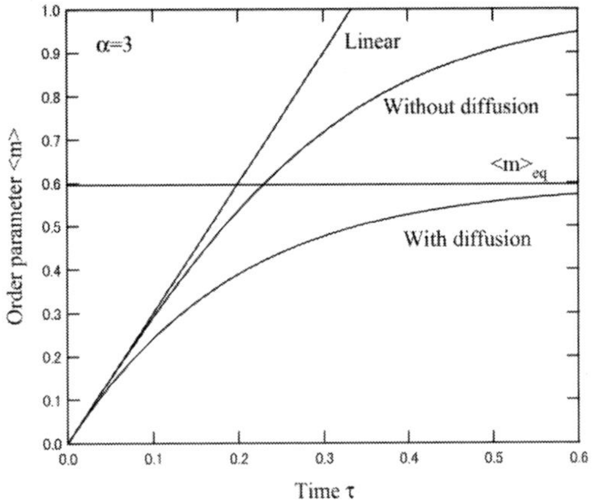

Figure 5. Comparison of the order parameter for the 2-dimensional magnetic alignment with/without rotational diffusion (α=3).

As the field parameter α is increased, the diffusion term in the RDE of Eq. (5) or (11) becomes much smaller than the drift one. As a consequence, the MA with rotational diffusion behaves like that without rotational diffusion. Finally, all of the above results are concerned with the positive α. When α is negative, the results are the same as above except that the order parameter is negative with the same absolute value.

In conclusion, we have made clear the influence of thermal disturbance on the dynamic process of the 2-dimensional magnetic alignment by obtaining the distribution function and the order parameter for the systems with/without rotational diffusion.

Acknowledgements

This work was supported by Grant-in-Aid for Scientific Research (Nos. 15085101 and 20560044) from MEXT of Japan and by JSPS Asian Program "Construction of World Center on Electromagnetic Processing of Materials".

References

[1] Yamaguchi M and Tanimoto Y ed. 2006 Magneto-science (Kodansha-Springer, Tokyo-Berlin)
[2] Fujiwara M, Oki E, Hamada M and Tanimoto Y 2001 *J. Phys. Chem.* A *Lett.* **105** 4383
[3] Kimura T, Yoshino M, Yamane T, Yamamoto M and Tobita M 2004 *Langumuir* **20** 5669
[4] Kohama T, Takeuchi H, Usui M, Akiyama J, Sung M G, Iwai K and Asai S 2007 *Mater. Trans.* **48** 2867
[5] Hirota N, Shimada T, Ando T, Wada H and Sakka Y 2007 *Int. Conf. on Magneto-Science* 3I-O4 (Hiroshima, Japan)
[6] Yamaguchi M, Ogawa K, Yamamoto I 2007 *Int. Conf. on Magneto-Science* 2O-11(Hiroshima, Japan)
[7] Doi M M and Edwards F 1986 *The Theory of Polymer Dynamics* (Oxford Univ. Press, Oxford)
[8] Kimura T, Sata H and Ito E 1998 *Polymer J.* **30** 455

Crystal Alignment of Primary Phase during Solidification

K Iwai[1], M Usui[2], N Toyoda[1] and S Asai[3]

[1] Department of Materials, Physics and Energy Engineering, Nagoya University, Furo-cho, Chikusa-ku, Nagoya, 464-8603, JAPAN

[2]IBIDEN Co.,LTD, 2-1Kanda-cho, Ogaki, 503-8604, JAPAN

[3]Innovation Plaza Tokai Japan Science and Technology Agency, 23-1 Ahara-cho, Minami-ku, Nagoya, 457-0063, JAPAN

E-mail: d42859a@cc.nagoya-u.ac.jp

Abstract. In this study, a crystal alignment method, in which imposing duration of a static magnetic field and an alternating current on a material during the solidification is controlled has been experimentally confirmed using a Sn-10% Pb metallic alloy. In the initial stage of the solidification both the magnetic and the alternating current are imposed on the alloy to break dendrites into pieces and to spread them to all the region in the sample. In the next stage, the static magnetic field is continuously imposed on the sample while the alternating current is turned off. The function of the static magnetic field in this stage is not only the rotation of the crystals to magnetically stable direction but also the suppression of the disturbance such as liquid motion. In the x-ray diffraction pattern of the sample solidified with the controlled imposition of the electromagnetic fields, peaks corresponding to theoretically predicted planes are mainly observed.

1. Introduction

A material with an anisotropic crystallographic structure shows anisotropy in physical properties such as electric conductivity, magnetic susceptibility, thermal conductivity and so on. That is, alignment of the crystals in the material enhances the anisotropy of these physical properties. Therefore, the crystal alignment is an attractive method for fabricating functional materials.

However, conventional methods for the crystal alignment have some problems. A mechanical method for the alignment of crystals such as a rolling has been industrially used [1], though the crystal alignment in this process is limited to a specific direction. Slip casting in a magnetic field [2] can be applied for the crystal alignment of ceramics. But this process is not suitable for large mass production because of low productivity. Epitaxial growth [3] is used as a thin film production process though formation of a bulk material is very difficult.

On the other hand, a static magnetic field has become a powerful tool for the crystal alignment under the development of the superconducting technology [4,5]. For the crystal alignment using a static magnetic field, environment in which crystals can easily rotate to magnetically stable direction is required. However, it is difficult to introduce this condition during solidification because a columnar solid usually forms from liquid phase [6]. Then, a new process for crystal alignment combining a rapid

Journal of Physics: Conference Series **156** (2009) 012008 doi:10.1088/1742-6596/156/1/012008

cooling and re-heating is proposed to introduce semi-liquid condition suitable for crystal rotation and the magnetic property of Bi-Mn alloy produced by this method was improved [7-9].

In this study, a crystal alignment method, in which imposing duration of a static magnetic field and an alternating current on a material during solidification is controlled has been experimentally confirmed using Sn-10% Pb metallic alloy.

2. Principle of Crystal Alignment

For crystal alignment using a magnetic field, anisotropy in magnetic susceptibility or that in a material shape is required. When a material with magnetic susceptibility, χ shape factor, N is submerged in a magnetic field, H, Magnetization energy, U is expressed as

$$U = -\frac{\mu_0 \chi}{2(1+\chi N)^2} H^2 \tag{1}$$

where μ_0 is magnetic susceptibility in vacuum.
In this equation, the term, χN is usually neglected because the magnetic susceptibility, χ is extremely smaller than unity and the shape factor is between 0 and 1. In this condition, the magnetization energy can be expressed as

$$U \approx -\frac{\mu_0 \chi}{2} H^2 \tag{2}$$

Therefore, a material with anisotropy in magnetic susceptibility should be chosen as a magnetic alignment material.

A Sn-10mass%Pb alloy has been adopted as a model alloy in this investigation. Because a primary phase at the liquidus temperature includes only 0.38atomic % of lead in mother tin as shown in Fig.1 and magnetic susceptibility of tin in a,b axis is larger than that in c axis [10], the primary phase might have the anisotropy in magnetic susceptibility. Under the imposition of a strong magnetic field, crystal planes perpendicular to the a-axis are expected to be perpendicular to the magnetic field direction.

For crystal alignment in a solidification process, not only the anisotropy in magnetic susceptibility but also other conditions are required. One is rotatable environment for primary phase particles to reduce the magnetization energy. Furthermore, the aligned particles must keep their direction against disturbances. In this investigation, the former can be achieved by imposing an alternating current and a static magnetic field in the initial stage of solidification. And the latter can be achieved by imposing a strong magnetic field.

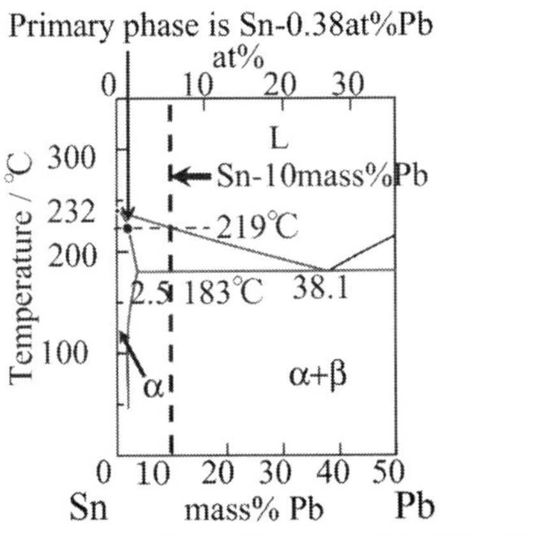

Figure 1. Phase diagram of Sn-Pb alloy

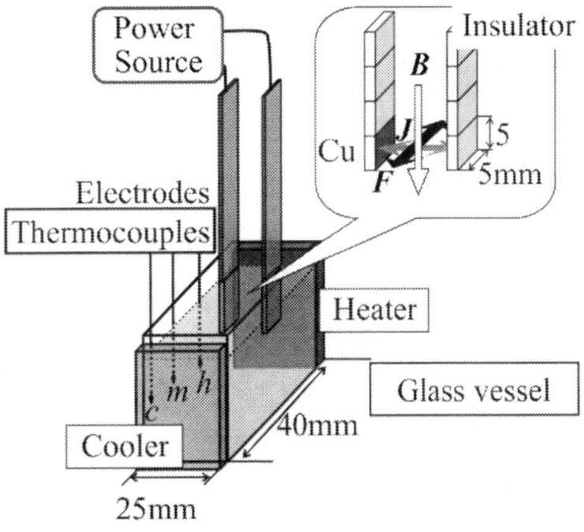

Figure 2. Experimental apparatus

3. Experiment

3.1. Experimental Setup

Experimental setup is shown in Fig.2. A molten sample of 300g was poured into a glass vessel with 25mm length and 40mm depth. One end of the vessel was cooled while the other end was heated to control the temperature distribution in the sample. A couple of copper electrodes was inserted into the molten sample in the vicinity of the heater to apply an alternating current to the sample. Temperatures in the sample were measured at three points by thermocouples. These points were located between the cooler and the heater along the long wall of the vessel as shown in Fig.2 (c : in the vicinity of cooler, h : in the vicinity of heater, m : middle position between them). This glass vessel was set in the bore of a super conducting magnet which generates a static magnetic field in the vertical direction. The Lorentz force is induced around the copper electrodes when the alternating current and the static magnetic field are simultaneously imposed on the sample.

We can introduce solid particles with the anisotropy in magnetic susceptibility by breaking dendrites around the electrodes where the Lorentz force is excited. These solid particles should exist not only around the electrodes but also in the other region for formation of the crystal aligned material.

3.2. Induction of convection by simultaneous imposition of alternating current and magnetic field

Convection is useful for spreading broken dendrites to the whole region in the sample. To confirm whether the convection is induced or not by the simultaneous imposition of the static magnetic field and the alternating current, we measured the temperature distribution in the molten sample during its cooling at the three points mentioned above. The measured temperature distribution in the sample is shown in Fig.3. Only the static magnetic field of 1tesla was imposed on the sample until the temperature in the vicinity of the cooler (c) decreased to 210C. The temperature difference in the sample was about 12K in this duration. When the temperature became 210C, the alternating current of 60A with frequency of 2kHz was turned on. Then, the temperature in the vicinity of the heater (h) decreased while that in the vicinity of the cooler (c) increased. And the temperature difference in the sample reduced to less than 4K. The measured temperatures in the sample oscillated with time. This was because that hot liquid lumps and cool liquid lumps circulated in the sample with slow thermal diffusion time. From this result, we can conclude that convection is induced by the simultaneous imposition of the static magnetic field and the alternating current on the sample.

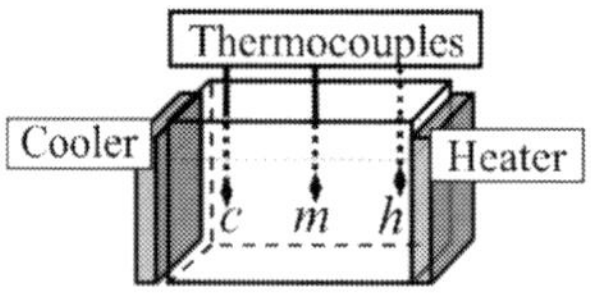

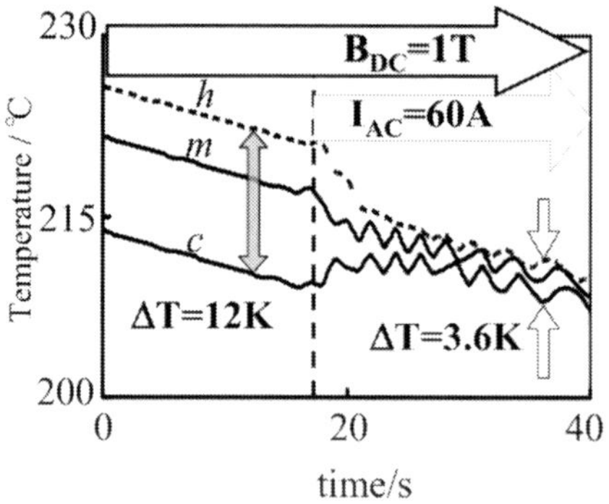

Figure 3. Effect of alternating current on temperature distribution in sample

3.3. Suppression of natural convection by magnetic field

Natural convection is induced by the temperature difference in the molten sample caused by the heater and cooler. This is disturbance from the viewpoint of the crystal alignment. Therefore, to confirm the effect of the magnetic field on the convection, we also measured the temperature distribution in the sample during its cooling with and without the imposition of the magnetic field. In the both cases, the electric current was not applied to the sample. The measured temperature distribution in the sample is shown in Fig.4. When the magnetic field was not imposed on the sample, the temperature difference in the sample was only about 1K. On the other hand, the temperature difference was larger than 10K under the imposition of the 7.5T magnetic field. This means that imposition of the 7.5T magnetic field has a function to suppress the natural convection in this experimental setup.

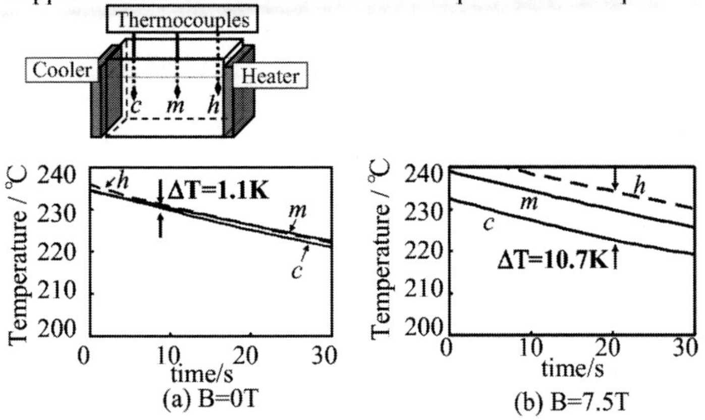

Figure 4. Effect of static magnetic field on temperature distribution in sample

3.4. Solidification experiment for crystal alignment

We solidified two samples. For the sample I, both the magnetic field of 7.5T and the alternating current of 80A, 2kHz were simultaneously imposed on the alloy to break dendrites into pieces and to spread them to the whole area of the sample in the initial stage of the solidification. In the next stage, only the static magnetic field was continuously imposed on the sample. The function of the static magnetic field in this stage was not only the rotation of the crystals to magnetically stable direction but also the suppression of the disturbance such as liquid motion. On the other hand, only the 6.5T magnetic field was imposed on the sample II during its solidification. After the solidification, the horizontal plane of the samples were cut and polished. Then, XRD analysis was done for these samples for the evaluation of the crystal alignment.

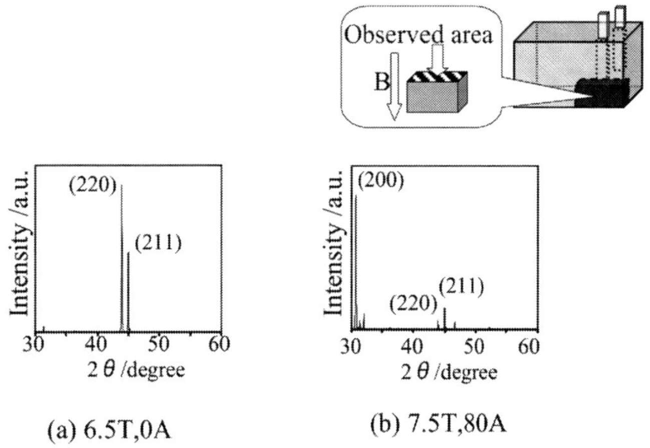

Figure 5. X-ray diffraction pattern

In the X-ray diffraction pattern of the sample solidified without the imposition of the alternating current, the first and second highest peaks were (220)-plane and (211)-plane of primary tin phase as shown in Fig.5 (a). Since the heat flow in this experiment was from the heater to the cooler under the suppression of convection, the preferable growing <110> direction of the primary phase must be parallel to this thermal flow direction [11]. This is the reason why the (220)-plane peak was the highest peak in the sample II. However, the number of grains in the XRD irradiated area might not be enough to deduce the result mentioned above since the thermal condition in this experiment satisfied columnar dendrite growing condition. On the other hand, the peaks corresponding to a,b-plane were mainly observed in the sample solidified under the imposition of the controlled electromagnetic field as shown in Fig.5 (b). In this sample the (211)-plane was suppressed in comparison with the sample II and the mainly observed peak of (200)-plane well agrees with the theoretical prediction. From these results, it is understood that the crystal alignment of the primary phase is achieved.

4. Conclusion

A crystal alignment method, in which imposing duration of a static magnetic field and an alternating current on a sample during the solidification is controlled has been experimentally confirmed using Sn-10% Pb metallic alloy.

Acknowledgement

This research was partially supported by JSPS Asian Core Program "Construction of the World Center on Electromagnetic Processing of Materials".

References

[1] Jee K K, Jangl W Y and Chung Y H 2003 J.Phys., IV France **112** 365
[2] Akiyama J, Hashimoto M, Takadama H, Nagata F, Yokogawa Y, Sassa K, Iwai K and Asai S 2005 *Materials Transactions JIM* **46** 203
[3] Watanabe T, Funakubo H, Saito K, Suzuki T, Fujimoto M, Osada M, Noguchi Y and Miyayama M 2002 *Applied Physics Letters* **81** 1660
[4] Sakka Y and Suzuki T 2005 *Journal of the Ceramic Society of Japan* **113** 22
[5] Kimura T, Maeda T, Sata H, Yamato M and Ito E 1995 *Polym.J.* **27** 247
[6] Yasuda H 2004 *Bull. Iron Steel Inst. Jpn.* **9** 789
[7] Yasuda H, Ohnaka I, Fukuda T, Watanabe K and Shimamura K 1999 *Proc. Japan-France Cooperative Science Seminar on New Aspect of Electromagnetic Processing of Materials* 88.
[8] Yasuda H, Ohnaka I, Yamamoto Y, Wismogroho A S, Takezawa N and Kishio K 2003 *Mater. Trans.* **43** 2550
[9] Yasuda H, Ohnaka I, Yamamoto Y, Tokieda Kand Kishio K 2003 *Mater. Trans.* **43** 2207
[10] Hoge H J 1935 *Physical Review* **48** 615
[11] Siqueira C A , Cheung N and Garcia A 2003 *Journal of Alloys and Compounds* **351** 126

Texturing hard Nd-Fe-B powdered ribbons in High Magnetic Field

S Rivoirard[1], V M T S Barthem[2], T Garcin[1], E Beaugnon[1], P E V de Miranda[3] and D Givord[4]

[1] CNRS/ Consortium de Recherche pour l'Emergence de Technologies Avancées (CRETA), BP 166, 38042, Grenoble Cedex 9, France

[2] UFRJ/Instituto de Física, CEP: 21941-972, Rio de Janeiro, RJ, Brazil

[3] UFRJ/COPPE, CEP: 21941-972, Rio de Janeiro, RJ, Brazil

[4] CNRS/Institut Néel, BP 166, 38042, Grenoble Cedex 9, France

E-mail: sophie.rivoirard@grenoble.cnrs.fr

Abstract. A new route to texture NdFeB alloys in magnetic field was presented in a recent paper [1], constituting an attempt towards the preparation of anisotropic bonded magnets. NdFeB ribbons composed of $Nd_2Fe_{14}B$ grains embedded in a Nd-Cu eutectic matrix, were annealed under an applied magnetic field up to 16T, at temperatures above the Nd-Cu melting temperature. A crystallographic texture was found to progressively develop at annealing temperatures, above 700 °C. In this paper, it is proposed that the grain orientation mechanism involves a competition between the aligning magnetic field torque acting on the magnetic grains and thermal disordering effects which becomes more and more significant as the temperature is increased. A simple model is developed to evaluate these effects. It is shown as well that a lowering of the alloy coercivity takes place during annealing.

1. Introduction

There exist two main types of NdFeB hard magnets. In the first type, the micron-size grains, obtained by grinding bulk ingots, have low coercive field values (typically, μ_0Hc is below 0.1 T). The large coercivity, which characterizes the magnets, develops only during sintering. In the second type, melt-spun amorphous Nd-Fe-B ribbons constitute the precursor material used for the preparation of bonded magnets. The sub-micrometer-size grains, formed during crystallization of melt-spun ribbons, are randomly oriented. Bonded magnets may be obtained by mixing ribbon flakes within a polymer. As a result of the crystallite random orientation, the magnetization remanence, μ_0M_r, is generally below 0.8T. Compared with these isotropic magnets, it may be hoped that anisotropic magnet powders with crystallographic alignment will lead to higher remanence, better rectangularity of the demagnetization curve, and, in turn, an increase in the energy product, $(BH)_{max}$.

Several routes have been explored with the aim of producing textured NdFeB hard nanostructured materials. This includes the development of a specific protocol of the HDDR process and various powdering processes applied to already textured bulk materials (die-upset ribbons or hot deformed alloys) [2-4]. The subtle effects involved in the texturing mechanisms have been described in recent reviews [5-7].

In permanent magnet alloys, texturing under magnetic field may be envisaged. In previous studies [8-9], magnetic texturing was tested at the alloy solidification temperature. The solidification of Sm-Co alloys under 7T led to a crystallographic preferential orientation of the c-axis (easy axis) of the solidified individual grains along the applied magnetic field direction. However, in Nd-Fe-B, the results showed that the magnetocrystalline anisotropy energy, in a magnetic field of 5 T, is not sufficient to align the $Nd_2Fe_{14}B$ crystallites, during their solidification, with the easy magnetization axis parallel to the field direction. In other studies, the crystallization of amorphous nano-composite magnets under magnetic field induced nanostructure refinement and better exchange coupling, but no significant texture effect was obtained [10,11].

In a recent study, we have demonstrated the possibility of orienting $Nd_2Fe_{14}B$ solid particles in an Nd-Cu liquid eutectic phase under a magnetic field applied on a ribbon heated above the melting temperature of the Nd-Cu eutectic phase [1]. In the present work, a quantitative analysis of the orientation mechanism in a magnetic field at high temperature is proposed, and the coercivity of field process samples is examined.

2. Experimental details

Ribbons were obtained by melt spinning an as-cast alloy of composition $Nd_{31.45}Fe_{66.34}B_{0.92}Cu_{1.31}$ equivalent to 50% wt $Nd_2Fe_{14}B$ and 50% wt $Nd_{70}Cu_{30}$. The set-up used for annealing in high magnetic field is described in [1]. The magnetic field, $\mu_0 H_{app}$, up to 16 T, was applied during the whole thermal treatment process consisting in two successive annealing steps.

The first step was aimed at orienting the particles under field. It was realized at temperatures T_1 above the melting temperature of the Nd-Cu eutectic, up to a maximum temperature of 1173 K. The plateau temperature was reached at a rate of 0.5 K/s and the duration of the temperature plateau was either 3 minutes or 1 hour. The second step, aiming at developing coercivity, was realized at 800 K for 30 minutes.

The samples were characterized by X-Ray Diffraction (not shown) and Scanning Electron Microscopy (SEM). The magnetic measurements were performed at room temperature, on bulk treated samples using a cryomagnet with 10 T maximum applied magnetic field. The macroscopic texture ratio, $r = (M_r^{//} - M_r^{\perp})/(M_r^{//} + M_r^{\perp})$ was taken as a measure of the degree of texture. With this definition, $0<r<1$, $\mu_0 M_r^{//}$ and $\mu_0 M_r^{\perp}$ are the remanent magnetizations measured in the parallel and perpendicular directions with respect to the direction of the applied field, H_{appl}.

The as-spun ribbons contained nano-sized $Nd_2Fe_{14}B$ crystallites embedded in the Nd-Cu eutectic matrix consisting of the β-Nd phase and the orthorhombic NdCu compound. A $Nd_2Fe_{14}B$ grain size of less than 10 nm was derived from the Scherrer formula. The melting temperature of the Nd-Cu eutectic is 783K [1]. Therefore, at temperature above 783 K, the $Nd_2Fe_{14}B$ grains are solid whereas the intergranular Nd-Cu phase is liquid. All thermal treatments reported in this work were performed under such conditions.

3. Results and discussion

The application of a magnetic field during annealing was found to have a remarkable effect on the alignment of the $Nd_2Fe_{14}B$ grains. As shown in Figure 1, an anisotropic magnetic behaviour is evidenced in the sample annealed at 1173K for 3 min, in a magnetic field of 16T. The direction parallel to the annealing field direction and in the ribbon plane is found to be the direction of preferred c-axis crystallite orientation. Figure 1 shows as well that the coercive field is significantly reduced upon annealing, from 2.5T in the un-annealed samples to 0.9T in the annealed ones. Additionally, in the sample treated in magnetic field, the coercivity decrease is more pronounced. After annealing at 1173K for 3min, the coercive field reaches 0.5T in the sample treated in 16T and 0.9T in the sample treated in zero field. This effect is attributed to a texture dependence of coercivity which is generally observed in this type of materials [12].

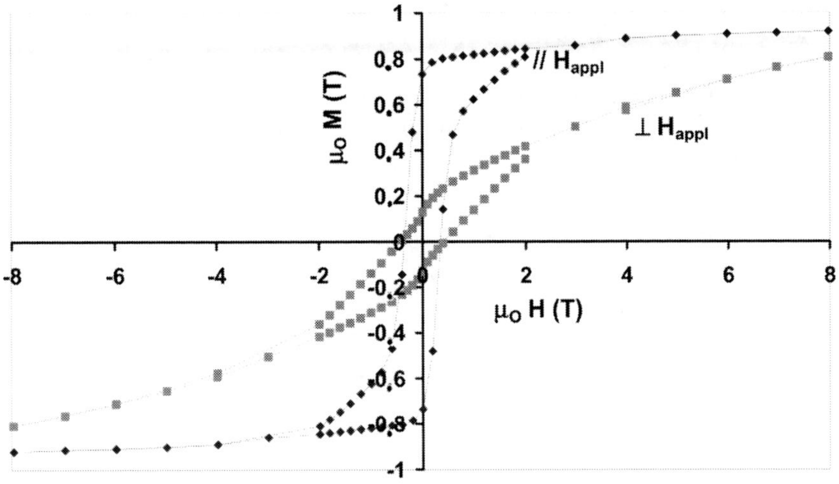

Figure 1. Room temperature magnetic behavior of the alloy treated at 1173 K, for 3 min in 16 T.

The field dependence of the texture ratio r measured after 3 min annealing at T_l=1173 K, is plotted in Figure 2; r is approximately proportional to H, it reaches 0.6 under the maximum applied field of 16 T. Note that for H=0T, r= -0.1, i.e. a weak basal plane texture is present [1].

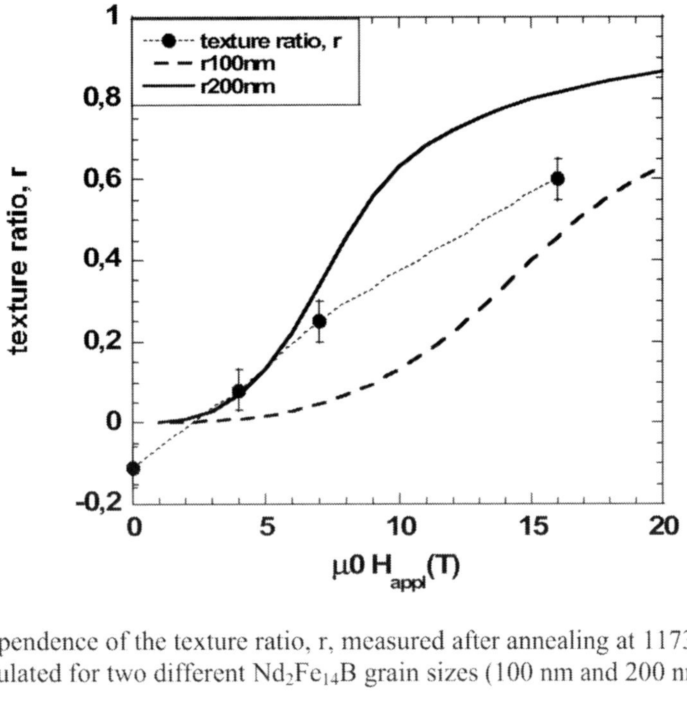

Figure 2. Field dependence of the texture ratio, r, measured after annealing at 1173 K for 3 min and calculated for two different $Nd_2Fe_{14}B$ grain sizes (100 nm and 200 nm).

The SEM picture of the same sample presented in figure 1 is shown in Figure 3. Coarse $Nd_2Fe_{14}B$ grains, several tens of microns in size, are observed. In other regions (grey regions in the Figure), nano-sized grains of the $Nd_2Fe_{14}B$ phase (typically 200 nm in diameter) embedded in the Cu-rich intergranular phase appear. This bimodal grain size distribution systematically shows up in all annealed samples. The $Nd_2Fe_{14}B$ crystallites appear to be essentially equiaxed and no particular shape alignment of the $Nd_2Fe_{14}B$ grains is found in the samples treated under magnetic field. No free iron or other soft magnetic phase is detected. The occurrence of large grains in the annealed samples may explain the low value of coercivity then observed. Indeed, it is well known that the coercivity of hard magnetic materials tend to decrease as the grain size is increased.

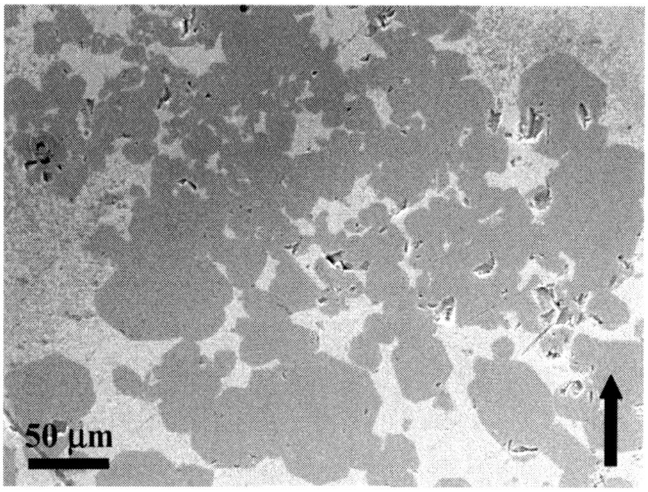

Figure 3. SEM pictures of the alloy treated at 1173 K, for 1 hour in 16 T (the arrow represents the field direction during annealing).

Within the mechanism proposed in the already cited paper [1], a liquid phase is required for efficient grain texturing. The texturing effect of the magnetic field occurs above the Curie temperature of the $Nd_2Fe_{14}B$ phase. The orientation of the tetragonal phase c-axes is determined by the magnetic field torque, Γ_{mag}. The friction torque within the liquid matrix is negligible, because of the low viscosity of the molten metal. Assuming that the magnetization of the considered $Nd_2Fe_{14}B$ grain is along the field, the magnetic torque, Γ_{mag}, may be expressed as

$$\Gamma_{magn} = K(T,H)\,V\sin(2\theta) \qquad (1)$$

where V is the grain volume, H_{appl} is the applied magnetic field, θ is the angle between the field and the easy direction and K(T, H_{appl}) is the magnetic anisotropy which, in the paramagnetic state, is a function of both T and H_{appl}. This equation shows that the magnetic torque is far much higher for the coarser grains than for the submicron size population of grains. It implies that, within the best experimental conditions (1173K and field up to 16T), the magnetic anisotropy is not the limiting factor for the coarse grains to orientate. Rather, friction between coarse grains will hinder their orientation, in regions where the amount of liquid phase is not sufficient. On the contrary, nanosized grains can rotate more easily but thermal activation may limit their orientation process.

For the purpose of discussing the respective role of both T and H_{appl} on the orientation mechanism of the small size grains, K(T, H_{appl}) has been calculated and a Boltzmann statistical calculation developed to numerically obtain the anisotropy ratio as a function of T and H_{appl}.

K(T, H_{appl}) has been evaluated based on simple considerations. In the tetragonal structure of the $Nd_2Fe_{14}B$ phase, second order anisotropy terms are expected to dominate and to first order, the anisotropy constant K is assumed to be proportional to μ_{Fe}^3 where μ_{Fe}^3 is the field induced Fe magnetic moment in the paramagnetic state expressed as a function of temperature within the molecular field model [1,13]. Thus, assuming $\theta \approx T_c$, (585 K) the $Nd_2Fe_{14}B$ paramagnetic Curie temperature, the anisotropy energy, $K(T, H_{appl})$ may be expressed as:

$$K(T,H_{appl}) = 3.2\times10^6 \left(\frac{\mu_0 H_{appl}}{T-\theta}\right)^3 \quad (2)$$

Calculated $K(T, H_{appl})$ values are presented in Figure 4 as a function of temperature and for different magnetic field intensities.

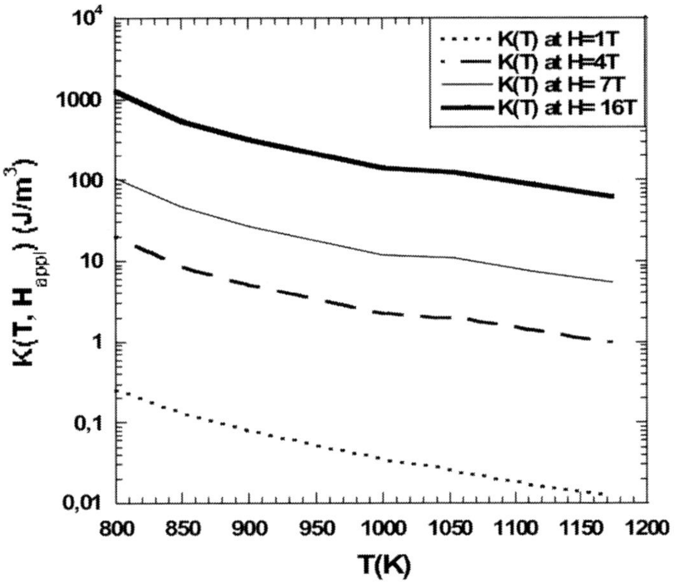

Figure 4. Anisotropy constant as a function of temperature for different magnetic field intensities.

The anisotropy constant value follows a logarithmic decrease as the temperature is increased. Moreover, a drastic increase of $K(T, H_{appl})$ is observed as the magnetic field is increased which implies, in turn, that the magnetic torque is much higher under high field. This means that magnetic orientation can be achieved for lower grain sizes under higher magnetic field.

Within Boltzmann statistics, it is assumed that the magnetization of all individual $Nd_2Fe_{14}B$ particles is saturated along the magnetic field. However, the crystallographic c-axes may be distributed about the field axis. Let θ be the angle between the applied field and the c-axis in a given grain. The texture ratio, r, is expressed as:

$$r = \overline{\cos\theta} = \frac{\int_0^{\pi/2} \sin\theta \cos\theta \, e^{\left(-\frac{K(T,H_{appl})V}{k_B T}\cos^2\theta\right)} d\theta}{\int_0^{\pi/2} \sin\theta \, e^{\left(-\frac{K(T,H_{appl})V}{k_B T}\cos^2\theta\right)} d\theta} \quad (3)$$

The texture ratio calculated as a function of the applied magnetic field for T=1173K and grains with 100 nm and 200 nm diameter is presented in Figure 2.

The calculated texture ratio for 200 nm diameter grains shows an upturn increase above 6T. At 12T the texture ratio reaches 0.75 and varies little under higher magnetic field. For 100 nm diameter grains, the effect of the magnetic field on texturing is significantly reduced. Under 16T, r is of the order of 0.5

only. These calculations show that thermal activation affects significantly the efficiency of grain texturing under field at this annealing temperature. However, calculations show that this degree of texture is already very significant, with $\overline{\cos\theta} = 0.85$ for r=0.45.

Globally, these results illustrate the fact that thermal disordering effects decrease as the grain size is increased, as a result of the increased particle whole anisotropy. A grain diameter of 200 nm is considered to be a minimum for satisfactory orientation to be achieved under the maximum available applied field of 16 T. Consequently, starting from grain sizes below 10nm in the as-cast ribbons, some grain coarsening is necessary as a first step for orientation to occur. Concomitantly, the number of grains which can orientate for a given magnetic field increases with magnetic field intensity. In the general case of magnet powders, where grain sizes are typically of the order of several hundred of nanometers, the high magnetic field texturing process may be very efficient.

The variation of the experimentally measured texture ratios lies between the curves calculated for 100nm and 200nm grain diameters. Actually, the experimental variation is in fair agreement with the calculation for 200 nm particles up to 7T. Such particle diameter corresponds approximately to the experimental diameter seen in the grey regions of Figure 3. At higher field values, the calculated texture ratio is above the experimental ones. This discrepancy may be attributed to the bimodal grain size distribution existing in the samples, whereas, a uniform grain size is considered in the calculation and coarse grains are not taken into account. The lower texture found experimentally with respect to the calculated ones suggests that orientation of the coarser grains is difficult. The same conclusion was reached in the initial study describing the effects of texturing NdFeB ribbons under field [1]. It was proposed that the difficulty in orienting large grains was due to the insufficient liquid phase amount which impeded free rotation of the grains under the effect of the aligning field.

4. Conclusion

An original approach towards the preparation of anisotropic hard magnetic alloys was explored in this work. Nd-Fe-B ribbons were annealed at various temperatures above the melting temperature of the liquid Nd-Cu eutectic. The application of a high magnetic field was found to result in a remarkable texturing effect. The crystallographic texture progressively developed with increasing the annealing temperatures, above 973 K. However, the alloy coercivity decreases concomitantly because of grain growth. The magnetic torque responsible for the texturing effect increases with increasing magnetic field intensity, leading to a texture ratio of 0.6 after annealing in 16T.

The calculation presented in this study indicates that grains diameters above 200 nm are required for satisfactory texturing to develop. At the same time, large micron-size grains are present in the alloy after annealing, which cannot orientate easily due to the fact that they come into contact during rotation. It is suggested that efficient orientation would require the use of alloys in which the amount of liquid phase is increased.

References

[1] Rivoirard S, Barthem V M T S, Bres R, Beaugnon E, de Miranda P E V and Givord D 2008 *J. Appl. Phys.* **104** 043915

[2] Mishima C, Hamada N, Mitarai H and Honkura Y 2000 *Proc. 16th Int. Workshop on Rare-Earth Magnets and their Applications* 873

[3] Lee R W 1985 *Appl. Phys. Lett.* **46** 790

[4] Rivoirard S, de Rango P, Fruchart D, Chastel Y and Martin C 2001 *Material Science and Engineering* A **311** 121

[5] Gutfleisch O, Khlopkov K, Teresiak A, Müller K H, Drazic G, Mishima C and Honkura Y 2003 *IEEE Trans. Magn.* **39** 2926

[6] Kirchner A, Hinz D, Panchanathan V, Gutfleisch O, Müller K H and Schultz L 2000 *IEEE Trans. Magn.* **36** 3288

[7] Rivoirard S and Givord D 2006 *Advanced Magnetic Nanostructur* (Springer) ed Sellmyer D and Skomski R chapter "hard magnetic nanostructures" p325

[8] Legrand B A, Chateigner D, Perrier de la Bâthie R and Tournier R 1998 *J. Alloys Compounds* **660-4** 275
[9] Courtois P, Perrier de la Bâthie R and Tournier R 1996 *J. Magn. Magn. Mater.* **153** 224
[10] Zhao T M, Hao Y Y, Hu X R, Yang Y S and Hu Z Q 1999 *J. Appl. Phys.* **85** 518
[11] Cui B Z, Yu C T, Han K, Liu J P, Garmestani H, Pechan M J and Schneider-Muntau H J 2005 *J. Appl. Phys.* **97** 10F308-1
[12] Givord D, Tenaud P and Viadieu T 1988 *J. Magn. Magn. Mat.* **72** 247
[13] Callen H B and Callen E 1966 *J. Phys. Chem. Solids.* **27** 1271

Journal of Physics: Conference Series **156** (2009) 012010 doi:10.1088/1742-6596/156/1/012010

In situ characterization of phase transformations in a magnetic field in Fe-Ni alloys

T Garcin, S Rivoirard and E Beaugnon

CNRS/CRETA 25 rue des martyrs BP166 38042 Grenoble CEDEX 9

E-mail: thomas.garcin@grenoble.cnrs.fr

Abstract. We have investigated the effect of magnetic field on the austenite(γ)-ferrite(α) equilibrium in Fe-xNi alloys with x = 0, 2, 4 wt%. The α→γ and γ→α transformations have been followed as a function of applied magnetic field by a laser dilatometer installed in a 16T superconducting magnet. In addition, magnetic measurements at high temperature have been used to follow the magnetic behavior of each alloy composition during a complete heat treatment.

We observe a shift of the phase diagram to higher temperature as the magnetic field is increased. We also find that the α-phase is either in the paramagnetic or in the ferromagnetic state as the transformation proceeds, depending on the amount of Ni. This results in an increase of the transformation temperature which is respectively proportional to the magnetic field if ferrite is formed in the ferromagnetic state and proportional to the square of the magnetic field if ferrite is paramagnetic.

1. Introduction

Modification of thermal treatments by an applied magnetic field in the processing of iron based alloys has been growing substantially in the last decade [1-15]. The goal of such processing is to achieve major improvements in material properties that cannot be obtained through conventional thermo-mechanical treatments. The relevant advances in magnet materials and design technologies can now provide magnetic fields of sufficient magnitude for new effect to be highlighted. Among these effects, modifications of phase stability [1-4], phase diagram boundaries [5-7], and transformation kinetics [8-10] have been observed when the involved phases exhibit different magnetization behaviors. Several papers report on theoretical approaches to study the impact of external magnetic fields on γ /α phase equilibria in Fe-C, Fe-X and Fe-C-X alloys [10-15]. These studies use the Weiss Molecular Field (WMF) assumptions [16] together with the Curie Weiss law to evaluate the change in the Gibbs Free energy of the individual phases and to calculate a field-modified phase diagram. All these studies result in a similar estimation of the magnetic field effect on the γ/α equilibrium temperature. According to them, the equilibrium temperature is raised from 1 to 3°C per Tesla depending on the alloy composition and magnetic field intensity.

Among various Fe based alloys, Fe-Ni alloys with low Ni contents are of special interest because the Curie temperature of the α phase (T_C) can be tuned by the alloy composition. Concomitantly, the γ→α and α→γ transformation temperatures are also changed. As an example, for Ni contents above 3%, the Curie point of the α phase becomes higher than the structural transformation temperature [17]. Therefore we can expect that the magnetic field dependence of the transformation temperature (T_T) will differ depending on the value of T_C-T_T. In the present work, we examine the α→γ and γ→α transformation temperatures in Fe-xNi alloys with x = 0, 2, 4 wt% by dilatometry measurements up to 16 T. In addition, magnetic measurements at high temperature are used to follow the magnetic behavior of each alloy composition during a complete heat treatment and to evaluate the magnetic state of ferrite during transformation.

2. Experimental procedure

Ingots of Fe-Ni alloys with content of 0, 2 and 4 wt% are prepared by induction melting in a cold crucible under argon atmosphere and cast into a 10 mm diameter copper mold. The alloys have been treated for 48h at 700°C under vacuum for homogenization. Finally the ingots have been cold worked down to a diameter of 4 mm and cut into cylindrical specimens of 10mm in length.

Transformation temperatures in magnetic fields have been measured by dilatation measurements. This home-made in-situ measurement setup, together with the furnace, is installed in the room temperature bore of a 16T superconducting magnet [19]. In this system, homogeneous temperature and field are applied to the 1 cm long sample placed in the magnet center region. The dilatation of the specimens is measured by a Michelson interferometer across an airtight pyrex window. The beam is reflected by a spherical mirror fixed on the top of an alumina sample holder.

Specimens have been heated up to the maximum temperature with a heating rate of 3°C/min and this temperature has been kept constant for 30 min. The same rate has been applied upon cooling. The maximum temperature has been set at 30°C above the upper boundary of the two phase $\alpha+\gamma$ region [17]. This correspond to 940°C for pure Fe, 900°C for Fe-2Ni wt% and 850°C for the Fe-4Ni wt% alloys.

The magnetic behavior of the involved phases has been determined by the Faraday's method using a magnetic balance. The measuring device together with the furnace are inserted in the vertical room temperature bore of a 7 Tesla superconducting magnet [19]. The sample temperature is measured with Pt/Pt-10%Rh thermocouples in direct contact with the sample. The maximum magnetic field obtained at the specimen location is about 4T with a magnetic field gradient of about 1.5T/cm.

The sample magnetization has been measured up to 1000°C with heating and cooling rates of 2°C/min. The magnetic field at the specimen location has been set to 2T with a magnetic field gradient of about 0.7T/cm.

3. Results

The dilatation signal of the Fe-2Ni wt% sample in a magnetic field of 16T is shown in Figure 1. The $\alpha\rightarrow\gamma$ phase transformation appears as a sharp contraction of the sample. The $\gamma\rightarrow\alpha$ phase transformation

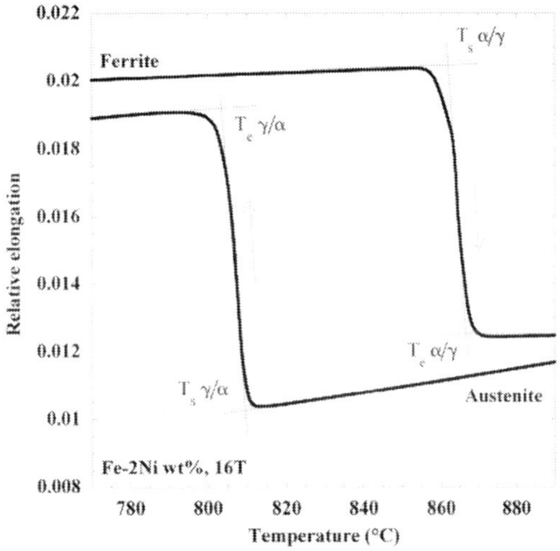

Figure 1. Dilatation measurement of Fe-2Ni wt% as a function of temperature in 16T.

exhibits a sharp dilatation. For each $\alpha\rightarrow\gamma$ and $\gamma\rightarrow\alpha$ transformation, the onset of the transformation temperature (T_S) and the end of the transformation temperature (T_E) are evaluated by the tangents method. The transformation temperatures (T_T) are defined as $T_{\alpha\rightarrow\gamma}$ on heating and $T_{\gamma\rightarrow\alpha}$ on cooling. They are calculated as the mean value of each T_E and T_S. The relative change of the transformation temperature due to the magnetic field (B), $\Delta T_{\alpha\rightarrow\gamma}$ and $\Delta T_{\gamma\rightarrow\alpha}$ are defined as:

$$\Delta T_{\alpha\rightarrow\gamma} = T_{\alpha\rightarrow\gamma}(B) - T_{\alpha\rightarrow\gamma}(0) \quad (1)$$
$$\Delta T_{\gamma\rightarrow\alpha} = T_{\gamma\rightarrow\alpha}(B) - T_{\gamma\rightarrow\alpha}(0) \quad (2)$$

As a reference, the transformation temperatures measured without any magnetic field $T_{\alpha\rightarrow\gamma}(0)$ and $T_{\gamma\rightarrow\alpha}(0)$ are given in Table 1.

Table 1. Measuring transformation temperatures on heating and cooling for Fe-Ni (0, 2 and 4 wt%) by dilatometry in zero field.

wt% Ni	0	2	4
$T_{\alpha\rightarrow\gamma}(0)$ °C	916	840	788
$T_{\gamma\rightarrow\alpha}(0)$ °C	903	775	695

The field dependences of $\Delta T_{\alpha\rightarrow\gamma}$ and $\Delta T_{\gamma\rightarrow\alpha}$ for the three alloys are plotted with solid marks in Figure 2(a) and 2(b) respectively.

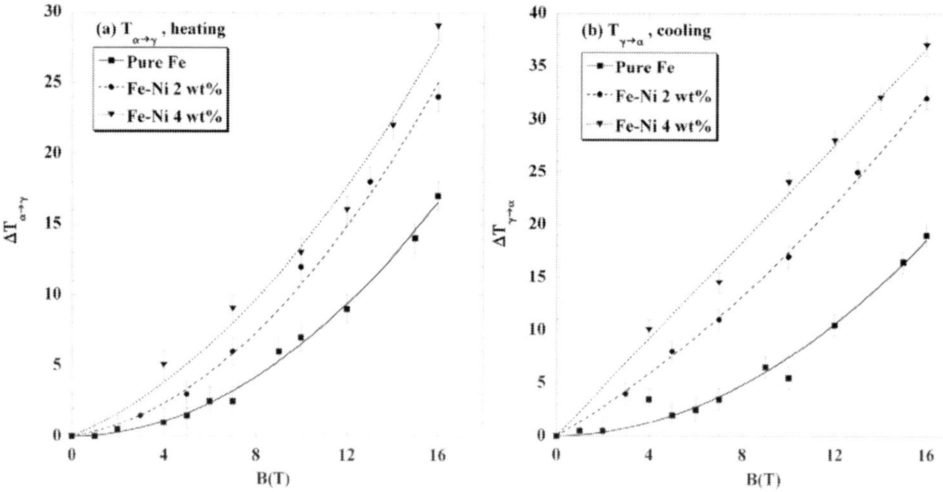

Figure 2. Change in the (a) $\alpha\rightarrow\gamma$ and (b) $\gamma\rightarrow\alpha$ transformation temperature by the application of magnetic field.

Obviously, $T_{\alpha\rightarrow\gamma}$ and $T_{\gamma\rightarrow\alpha}$ increase with increasing the magnetic field intensity for the three alloys. The increase of $T_{\alpha\rightarrow\gamma}$ amounts to 17°C in pure Fe, 24°C in the Fe-2Ni wt% alloy and 29°C in the Fe-4Ni wt% alloy in a magnetic field of 16T. This increase seems to be proportional to B^2 for the three alloys compositions. A similar increase is observed for the $T_{\gamma\rightarrow\alpha}$ temperature. This increase seems to be also proportional to B^2 for the pure iron and the 2wt% Ni alloys composition, whereas it tends towards a linear behavior for the 4wt% Ni alloy.

Figure 3 shows the magnetization curves as a function of the temperature obtained for the three alloys upon heating in Figure 3(a) and cooling in Figure 3(b). Upon heating, a sharp decrease in the magnetic signals is associated to the respective Curie points of pure iron (770°C), Fe-2Ni wt% (764°C) and Fe-4Ni wt% (759°C) given by [13]. At higher temperature, the $\alpha \rightarrow \gamma$ transformation is observed as a small decrease in the magnetic signal. This effect can hardly be seen in the Fe-4Ni wt% alloy because the magnetic transition occurs in the same temperature range as the $\alpha \rightarrow \gamma$ transformation. Upon cooling, the $\gamma \rightarrow \alpha$ transformations and the magnetic transitions are well dissociated in pure Fe. In the Fe-2Ni wt% and Fe-4Ni wt% alloy composition, ferrite appears in its ferromagnetic state thus leading to a sharp increase in the magnetic signal.

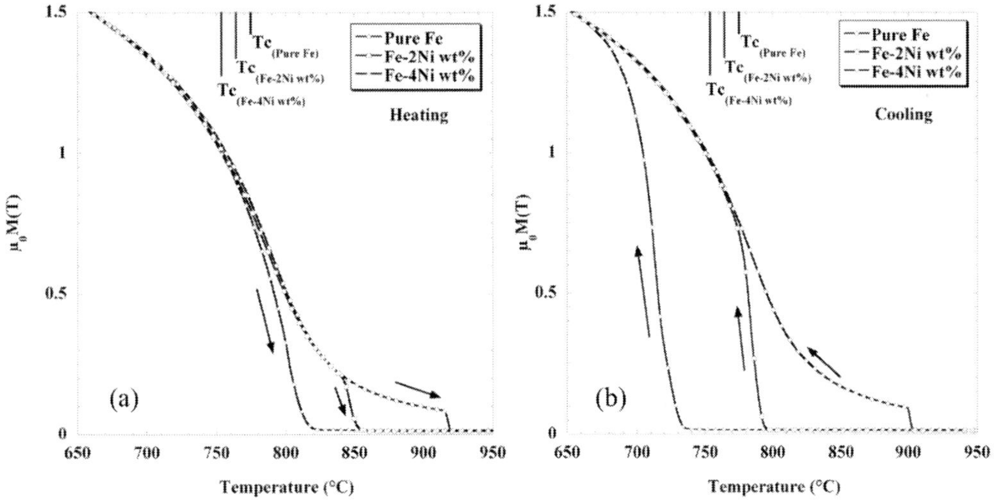

Figure 3. Magnetization curves as a function of temperature at 2°C/min for Fe-Ni (0, 2 and 4 wt%), upon (a) heating and upon (b) cooling.

4. Discussion

The effect of a high magnetic field on the respective stability of the ferrite (α) and austenite (γ) phases can be qualitatively explained in term of the magnetic contribution to the total Gibbs energy of each phase. When a magnetic induction, B, is applied, the respective Gibbs free energy of ferrite and austenite are lowered due to the contribution of a magnetic energy term, ΔE_m, to the total energy of each phase:

$$\Delta E_m = -\int_0^{B_{appl}} M\, dB \qquad (3)$$

where M is the field dependant magnetization projected along B. These respective contributions are proportional to the field dependence of the magnetization of each phase. As ferrite is more magnetic than austenite, this phase is stabilized, so that a shift in the equilibrium temperature is observed towards higher temperatures. Moreover, the magnetic contribution of the austenite phase to the Gibbs free energy can be neglected due to the very low magnetization of this paramagnetic phase.

The shift of the $\gamma \rightarrow \alpha$ transformation temperature was evidenced for all the alloy compositions studied in this work. However and depending on the Ni content, the variation of the equilibrium temperature shows two different behaviors with the magnetic field intensity: the transformation temperature increases either linearly or follows a quadratic behavior with the applied magnetic field. These different variations are linked with the Curie temperature of the considered compound. When the $\gamma \rightarrow \alpha$ transition occurs below the Curie temperature of ferrite, the magnetization of the ferrite phase, M, is rapidly saturated so that the magnetic energy is roughly linear with the applied magnetic field value. On the contrary, when the $\gamma \rightarrow \alpha$ transition occurs above the Curie temperature, the ferrite is formed in the paramagnetic state and hence M is almost linearly dependent on B. Since the magnetic energy is proportional both to the applied field and to the magnetization, the temperature shift is roughly proportional to the square of the field value. In this particular case, a more accurate analysis should use the Weiss Molecular Field model as the transformation occurs close to the Curie temperature of the ferrite phase. These two different behaviors have been recently reported in steels and cast irons [19] as well as in Fe-Co alloys [21].

5. Conclusion

In this work, a new experimental method based on in-situ measurements has been used to monitor the $\alpha \rightarrow \gamma$ and $\gamma \rightarrow \alpha$ transformations in Fe-Ni alloys as a function of applied magnetic field intensity. The transformation temperature is shifted towards higher temperatures in a magnetic field. This shift follows either a linear or quadratic behavior with the magnetic field intensity. Experimental determination of the magnetization of the ferrite phase at high temperature in various field values is in progress and will allow direct comparisons between WMF predictions and experiments.

References

[1] Enomoto M, Guo H, Tazuke Y, Abe Y R, Shimotomai M 2001 *Metall. Mater. Trans.* **A32** 445
[2] Hao X, Ohtsuka H 2004 *Mater. Trans.* **45**(8) 2622
[3] Kakeshita T, Fukuda T 2006 *Sci. Technol. Adv. Mater.* **7** 350
[4] Rivoirard S, Garcin T, Gaucherand F, Bouaziz O, Beaugnon E 2006 *J.of Physics.* **51** 541
[5] Zhang Y, Esling C, Lecomte J S, He C S, Zhao X, Zuo L 2005 *Acta Mater.* **53** 5213
[6] Maruta K, Shimotomai M 2002 *J. of Crystal Growth* **237** 239 1802
[7] Shimotomai M, Maruta K, Mine K, Matsui M 2003 *Acta Mater.* **51** 2921
[8] Ludtka G M, Jaramillo R A, Kisner R A, Nicholson D M, Wilgen J B, Mackiewicz-Ludtka G, Kalu P N 2004 *Scripta Mater.* **51** 171
[9] Zhang Y, He C, Zhao X, Zuo L, Esling C 2005 *J Magn. Magn. Mater.* **294** 267
[10] Rivoirard S, Gaucherand F, Bouaziz O, Pinto da Costa E, Beaugnon E 2006 *ISIJ Int.* **46**(9) 1274
[11] Joo H D, Kim S U, Shim N S, Koo Y M 2000 *Mater. Letters* **43** 225
[12] Gao M C, Bennet T A, Rollet A D, Laughlin D E 2006 *J. Physics D* **39** 2890
[13] Guo H, Enomoto M 2000 *Mater. Trans.* JIM **41**(8) 911
[14] Choi J K, Ohtsuka H, Xu Y, Choo W Y 2000 *Scripta Mater.* **43** 221
[15] Zhang Y, Esling C, Calcagnotto M, Gong M L, Zhao X, Zuo L 2007 *J. Physics D* **40** 6501
[16] R M Bozorth *Ferromagnetism* 1978 IEEE Press, The Institute of electrical and electronics Engineers, Inc New York 430
[17] Massalski T B *Binary alloy phase diagrams, second edition* 1990 **2** 1737
[18] Ishikawa Y *Handbook of Magnetic substance* 1975 ed. by Chikazumi S et al. Asakura Shoten, Tokyo 289
[19] Gaucherand F, Beaugnon E, *Phys. B* 2001; **294** 96
[20] Beaugnon E, Rivoirard S, Garcin T, Bouaziz O 2006 *Proc. 5th Int. Symp. on Electromagn. Process of Mater.* 604
[21] Fukuda T, Yuge M, Terai T, Kakeshita T 2006 *J. of Physics 2006* **51** 307

Journal of Physics: Conference Series **156** (2009) 012011 doi:10.1088/1742-6596/156/1/012011

Selective formation of an ordered variant in $Fe_{45}Pd_{55}$ alloy under magnetic field

S Farjami, T Fukuda and T Kakeshita

Department of Materials Science and Engineering, Graduate School of Engineering, Osaka University, 2-1, Yamada-Oka, Suita, Osaka 565-0871, Japan

Email: farjami@mat.eng.osaka-u.ac.jp

Abstract. The effect of magnetic field on microstructure control has been investigated during disorder-order transformation in a bulk single crystal of $Fe_{45}Pd_{55}$ alloy, where a two-step ordering heat-treatment has been made to clarify the effect of magnetic field. As a result, single variant of $Fe_{45}Pd_{55}$ is obtained when a magnetic field of 10 T is applied only at the early stage of ordering heat-treatment, suggesting that magnetic field is especially effective at the early stage of ordering process.

1. Introduction

Near-equiatomic Fe-Pd alloys undergo a cubic (A1) to tetragonal ($L1_0$) transformation and the ordered tetragonal phase is known to exhibit a large magnetocrystalline anisotropy with an "easy" *c*-axis [1-3]. In the ordered phase, three crystallographic domains (variants) form equivalently in the absence of external fields. However, under a magnetic field, there arises a difference in magnetic energy among the variants. Therefore, we can expect selective formation of an ordered variant with the lowest magnetic energy under a magnetic field.

In fact, formation of a single variant with its easy axis aligned to the applied magnetic field during disorder-order transformation was firstly reported in Fe-50at%Pd [4-5]. However, how and in which stage magnetic field is effective for the selective formation of specific variant has not been clarified yet. Recently, we have revealed by using a Co-50at%Pt alloy that a single variant of ordered $L1_0$-phase can be obtained by applying magnetic field only in the early stage of ordering heat-treatment [6]. These results suggest that magnetic field is especially effective at the early stage of ordering. Considering the similarity of crystal structure and magnetic properties between CoPt and FePd, we can expect that the single variant of $L1_0$-type FePd is also obtained by applying magnetic field only in the early stage of ordering heat-treatment.

In the present study, therefore, we have further examined the effect of magnetic field on selective formation of an ordered variant in $Fe_{45}Pd_{55}$ alloy by a two-step ordering heat-treatment. The first step of ordering is made under a magnetic field applied in the [001] direction of disordered specimen followed by the second step of ordering at higher temperature under no magnetic field to complete the disorder-order transformation.

2. Experimental procedure

An ingot of Fe-Pd alloy was prepared by arc melting in an argon atmosphere with a nominal concentration of 45at% Fe and 55at% Pd. The ingot was grown into a single crystal by a floating zone method. The single crystalline rod was homogenized at 1273 K for 168 h and then quenched in iced-water. Some specimens with all surfaces parallel to {001} planes were cut out for ordering heat-treatment and electrical resistivity measurement as shown in figure 1. Three edge directions of cube specimens are defined as X-, Y-, and Z-directions. After cutting the specimens into suitable shapes for each measurement, they were heat-treated at 1173 K for 1 h followed by iced-water quenching to obtain the disordered state.

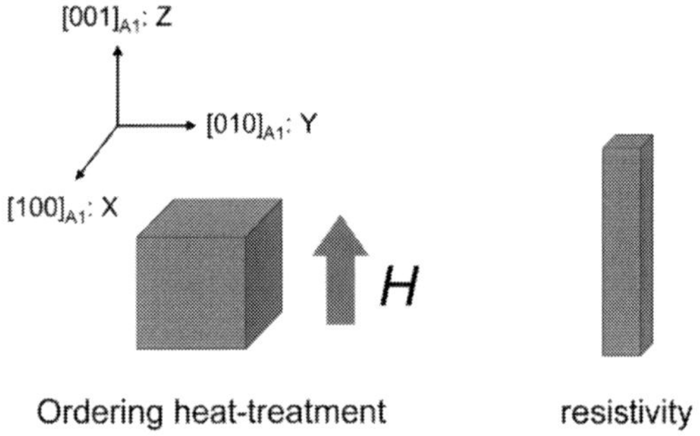

Figure 1. Schematic illustrations of cube and rectangular parallelepiped specimens prepared for ordering heat-treatment under magnetic field of *H* and electrical resistivity measurement.

In order to know the influence of magnetic field on the microstructure formation clearly, we have made a two-step ordering heat-treatment as shown schematically in figure 2. The first step of ordering is made under a magnetic field of 10 T applied in the $[001]_{A1}$ direction to initiate the ordering under the effect of magnetic field. Then it is followed by the second step of ordering under no magnetic field to complete the ordering process.

In this study, the variant whose easy axis lies in the Z-direction is expressed as Z-variant. The other two variants (X-, Y-variants) are defined in the same way. After the first and second step of ordering, magnetization of the cube specimens was measured at 300 K along the X-, Y-, and Z-directions by a superconducting quantum interference device (SQUID) magnetometer to know the fraction of the variants. To determine the ordering temperature of the two-step ordering heat-treatment, resistivity was measured in heating and cooling processes in disordered and ordered specimens by a direct current four-probe method.

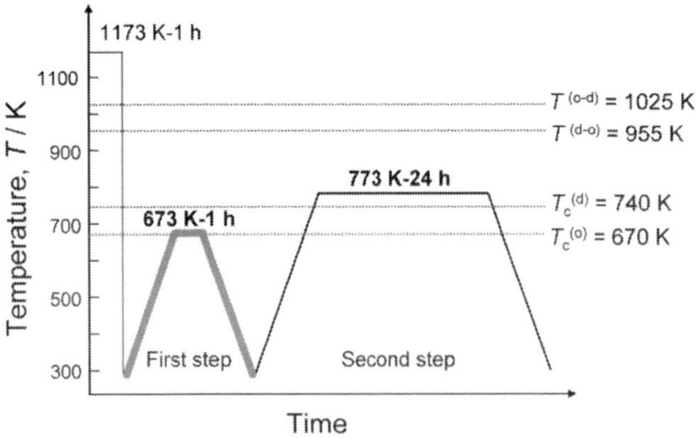

Figure 2. Schematic diagram of the two-step ordering heat-treatment. The bold lines indicate the ordering heat-treatment under a magnetic field and the thin lines indicate the ordering heat-treatment under no magnetic field.

3. Results

Temperature dependence of resistivity during heating and cooling processes of disordered and ordered specimens is shown in figure 3. The dotted curve is resistivity of disordered specimen measured with a heating rate of 100 K/min. The resistivity curve (dotted curve) exhibits a bending at 740 K which corresponds to the Curie temperature of the disordered phase, $T_c^{(d)}$. The solid curve is resistivity of the ordered specimen measured in the heating process (1 K/min) after a preliminary ordering heat-treatment at 773 K for 24 h. The resistivity heating curve (solid curve) exhibits a bending at 670 K which corresponds to the Curie temperature of the ordered phase, $T_c^{(o)}$. On further heating, the resistivity increases gradually and then decreases sharply due to the transition from the ordered phase to the disordered phase. The order-disorder transformation temperature, $T^{(o\text{-}d)}$ is about 1025 K. The dashed curve is corresponding cooling curve measured from 1200 K to room temperature with a cooling rate of 1 K/min. The disorder-order transformation temperature, $T^{(d\text{-}o)}$ is shown as a sharp increase in resistivity cooling curve (dashed curve) at 955 K. The transformation temperatures are shown with vertical arrows on each curve.

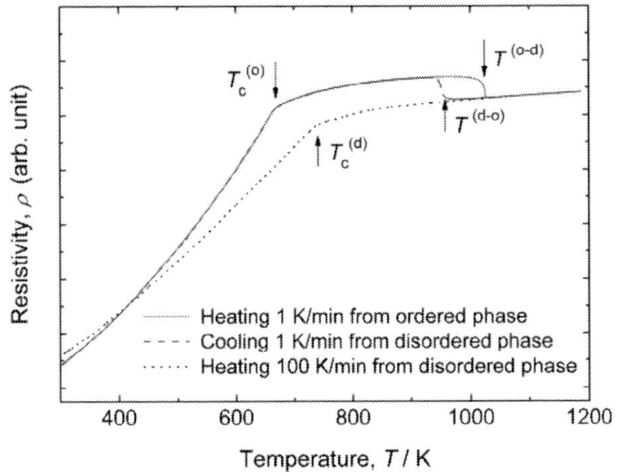

Figure 3. Temperature dependence of electrical resistivity.

Considering the result of resistivity measurement, the conditions of the two-step ordering heat-treatment have been determined: The first step of ordering at 673 K, which is close to the Curie temperature, for 1 h under a magnetic field applied in the [001] direction (Z-direction), followed by the successive second step of ordering at 773 K for 24 h. These temperatures are indicated in figure 2.

Magnetization curves measured at 300 K along the X-, Y-, and Z-directions of as-quenched specimen is shown in figure 4(a). There is no remarkable difference among the curves and they saturate in a low field suggesting that a nearly disordered state is obtained by quenching from 1173 K. Figure 4(b) shows magnetization curves measured after the first step of ordering at 673 K for 1 h under 10 T field. The figure indicates the progress of ordering by this heat-treatment and fraction of the variant whose easy axis lies in the applied field direction (Z-variant) is larger than the fraction of the other two variants (X-, and Y-variants) as shown clearly in the inset of figure 4(b). Magnetization curves measured along the X-, Y-, and Z-directions after the second step of ordering at 773 K for 24 h under no magnetic field are shown in figure 4(c). From the figure it is clear that a single variant of $Fe_{45}Pd_{55}$ alloy is obtained.

Similar ordering heat-treatment has been made without applying a magnetic field to assure that selective formation of specific variant is caused by magnetic field. As a result, we have experimentally confirmed that the three variants form nearly equivalently when the first step of ordering is made in the absence of magnetic field.

The result of the two-step ordering heat-treatment described above suggests that selection of variant by magnetic field is essentially determined in an early stage of ordering. That is, during the early stage of ordering in the first step, the fraction of the Z-variant reaches some specific amounts larger than the fraction of the X- and Y-variants due to the application of magnetic field. When this initial condition is satisfied, magnetic field is not necessary in the following ordering process. For further discussion we need to clarify the microstructure formed at the early stage of ordering under a magnetic field and it is a subject in the future.

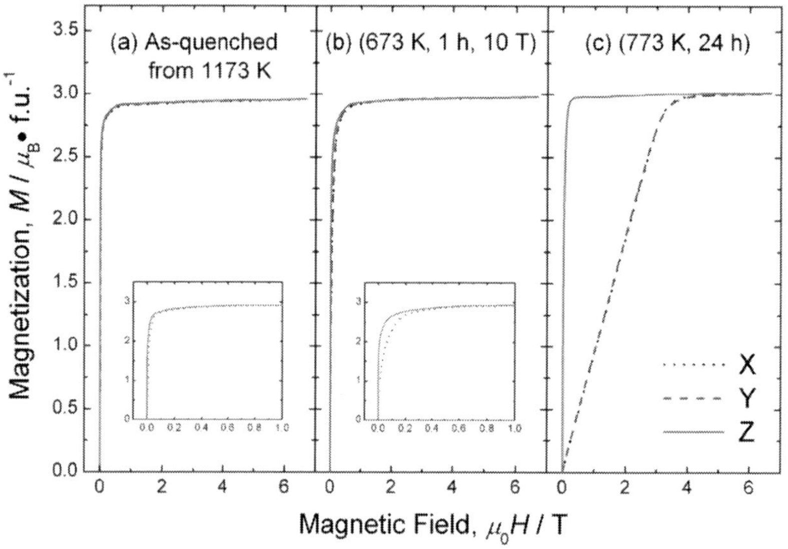

Figure 4. Magnetization curves measured at 300 K along the X-, Y-, and Z-directions of (a) as-quenched specimen, (b) after the first step of ordering at 673 K for 1 h under magnetic field of 10 T, (c) successive second step of ordering at 773 K for 24 h.

4. Conclusions

We have investigated effect of magnetic field on microstructure formation through disorder (A1)-order ($L1_0$) transformation by using single crystal of $Fe_{45}Pd_{55}$ alloy. A single variant is realized when the ordering heat-treatment is made close to the Curie temperature under magnetic field of 10 T applied in the [001] direction of the disordered phase. These results suggest that magnetic field is effective at the early stage of ordering process, as in CoPt previously studied by the authors [6].

Acknowledgements

The authors gratefully acknowledge Professor S. Nishijima of Osaka University for providing the superconducting magnet throughout the course of this work. This study was supported by Priority Assistance for the Formation of Worldwide Renowned Centers of Research–The Global COE Program (Project: Center of Excellence for Advanced Structural and Functional Materials Design) from the Ministry of Education, Culture, Sports, Science and Technology (MEXT), Japan.

References

[1] Gushchin G M and Berseneva F N 1987 *Fiz. Met. Metalloved.* **63** 926
[2] Zhang B, Lelovic M and Soffa W A 1991 *Scr. Metall. Mater.* **25** 1577
[3] Klemmer T, Hoydick D, Okumura H, Zhang B and Soffa W A 1995 *Scr. Metall. Mater.* **33** 1793
[4] Oshima R and Tokoro H 1998 *J. Japan Inst. Metals* **62** 317
[5] Tanaka K, Ichitsubo T, Amano M, Koiwa M and Watanabe K 2000 *Mater. Trans. JIM* **41** 917
[6] Farjami S, Yasui M, Fukuda T and Kakeshita T 2008 *Scr. Mater.* **58** 811

Effect of magnetic field on solid-solid phase transformations in iron-based ferromagnetic alloys

T Kakeshita and T Fukuda

Department of materials Science and Engineering, Graduate School of Engineering, Osaka University, 2-1, Yamada-oka, Suita, Osaka 565-0871, Japan

E-mail: kakeshita@mat.eng.osaka-u.ac.jp

Abstract. We have investigated the effect of magnetic field on phase transformation of iron-based ferromagnetic alloys. Two systems are selected. One is Fe-Rh alloys exhibiting a diffusional transformation from an austenite phase to a ferrite phase on cooling. The field dependence of the transformation temperature is well explained thermodynamically. The other is an Fe-31.2Pd alloy exhibiting a martensitic transformation. In this system, rearrangement of martensite variants occurs by the application of the magnetic field, and this behaviour is quantitatively explained by considering magnetic shear stress, which is expressed by the magnetic energy difference between variants divided by twinning shear.

1. Introduction

Solid-solid phase transformations are usually classified into two groups from a viewpoint of atom diffusion: diffusionless (martensitic) transformation and diffusional transformation. For both transformations, their characteristics are known to be influenced by external fields such as temperature, hydrostatic pressure and uniaxial stress. Magnetic field is one of such fields, and especially effective for changing the transformation temperature when there is a large difference in magnetization between the high- and low-temperature phases. In fact, concerning diffusionless (martensitic) transformations, the effects of magnetic field have been intensively studied in some iron based alloys[1-3]. Through the studies, the field-dependence of transformation temperature and also transformation kinetics have been quantitatively clarified until now. On the other hand, concerning diffusional transformations, the effect of magnetic field has not been so clarified yet.

γ(austenite)$\leftrightarrow\alpha$(ferrite) transformation in Fe-based alloys is one of representative diffusional transformations, in which the transformation temperature is considerably influenced by magnetic field[4-6], because there exists a large difference in magnetization between the two phases. Among various Fe-based alloys, Fe-Rh system is especially of interest because the Curie temperature T_c of the α-phase becomes higher than the $\gamma\leftrightarrow\alpha$ transformation temperatures when Rh content exceeds a certain value while the relation reverses when the content is below it. Then we can expect that the field-dependence of $\gamma\leftrightarrow\alpha$ transformation temperatures in the former case differs from that in the latter case.

In the present study, therefore, we have examined the $\gamma\rightarrow\alpha$ and $\alpha\rightarrow\gamma$ transformation temperatures in Fe-xRh alloys with x = 2, 5, 10 at.% under various magnetic field by electrical resistivity measurement. We have also measured magnetizations of the α- and γ-phases near the transformation temperature as well as the latent heats of the transformations. Then the field-dependence of the $\gamma \rightarrow \alpha$ and $\alpha \rightarrow \gamma$ transformation temperatures is discussed on the basis of a Clausius-Clapeyron-like equation.

In addition, we will show rearrangement of variants by magnetic field observed in Fe-31.2Pd alloy, and make a quantitative analysis on this bahavior by introducing a magnetic shear stress.

2. Effect of magnetic field on γ-α transformation of Fe-Rh alloys

2.1. Sample preparation

Ingots of Fe-Rh alloys with Rh content of 2, 5, 10 at.% were prepared by arc melting. They were homogenized in vacuum for 72 h at 1273 K followed by quenching into iced-water. From each ingot, specimens for electrical resistivity measurement, magnetization measurement and differential scanning calorimetry (DSC) were cut out. The size of the specimen is 20 mm × 2 mm × 1 mm for electrical resistivity measurement, 5 mm × 3 mm × 1 mm for magnetization measurement, and 3 mm × 3 mm × 1 mm for DSC measurement. After cutting out these specimens, they were subjected to a heat-treatment for 1 h at 1273 K.

2.2. Transformation temperatures under magnetic field

Figure 1(a) shows temperature dependence of electrical resistivity of the Fe-2Rh alloy, and figure 2 shows that of the Fe-5Rh and Fe-10Rh alloys. Measurements have been made under magnetic field of 10 T and without applying magnetic field. The heating and cooling rate is 1 K/min, and the resistivity curve under magnetic field of 10 T is shifted upward by a constant value of β = 0.3 μΩm to separate the curves.

First, we determine the Curie temperature T_c of the α-phase from the resistivity curves. The heating and cooling curves of the Fe-2Rh alloy measured under the field strength of $\mu_0 H$ = 0 T (without magnetic field) exhibits a bend at about 1036 K, which is indicated by a double arrow. The bend point disappears when magnetic field of 10 T is applied as seen in figure 1(a). Considering the fact that the Curie temperature disappears under the magnetic field, the bend point corresponds to T_c of the α-phase. The heating curve of the Fe-5Rh alloy (figure 2(a)) also exhibits a bend point at 1029 K as indicated by a double arrow, and the bend point disappears when the magnetic field is applied. From the result, T_c of the Fe-5Rh alloy is determined to be 1029 K. In case of the Fe-10Rh alloy, T_c of the α-phase is not detected by resistivity measurement because the α-phase transforms to the γ-phase on heating before T_c is reached. Thus, we have roughly estimated T_c of the Fe-10Rh alloy by extrapolating T_c's of pure Fe (1044 K)[7], Fe-2Rh and Fe-5Rh alloys. The estimated value is about 1015 K, which will be used for a thermodynamic analysis later. Concerning the γ-phase of the present alloys, the Curie temperature is expected to be far below the $\gamma \rightarrow \alpha$ transformation temperature. Thus γ-phase of the present alloys is paramagnetic in the examined temperature range.

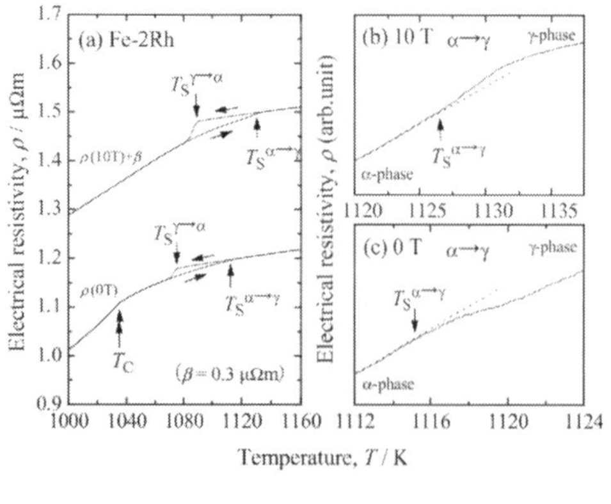

Figure 1. Temperature dependence of electrical resistivity of a Fe-2Rh alloy measured under magnetic field of 10 T and without magnetic field. (b), and (c) show the $\alpha \rightarrow \gamma$ transformation start temperature clearly as a deviation from linear relation (dotted lines) under magnetic field of 10 T and without magnetic field.

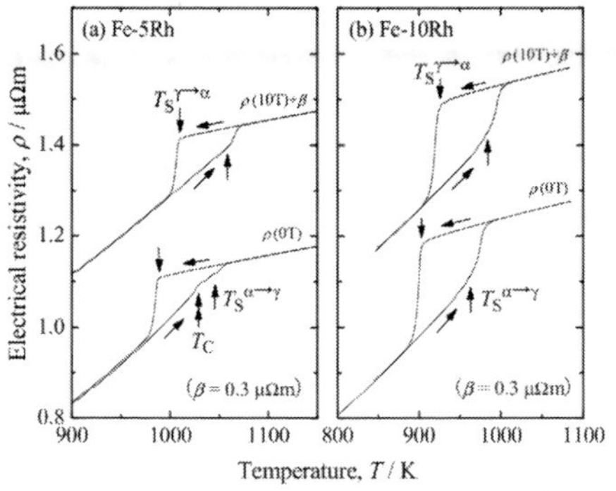

Figure 2. Temperature dependence of electrical resistivity of (a) Fe-5Rh alloy, and (b) Fe-10Rh alloy under magnetic field of 10 T and without magnetic field.

Next, we determine the $\gamma \rightarrow \alpha$ transformation temperature $T_s^{\gamma \rightarrow \alpha}$. This transformation is detected as a sharp decrease in resistivity for all the alloys, and we show $T_s^{\gamma \rightarrow \alpha}$ with an arrow on each cooling curve. Obviously, $T_s^{\gamma \rightarrow \alpha}$ under magnetic field of 10 T is higher than that under no magnetic field for each alloy.

Finally, we determine the $\alpha \rightarrow \gamma$ transformation temperature $T_s^{\alpha \rightarrow \gamma}$. The resistivity heating curves of the Fe-5Rh and Fe-10Rh alloys show a sharp increase in association with the $\alpha \rightarrow \gamma$ transformation. Thus, $T_s^{\alpha \rightarrow \gamma}$ is easily determined as indicated by arrows in figure 2 (a) and (b). In case of the Fe-2Rh alloy, however, the $\alpha \rightarrow \gamma$ transformation does not appear as a sharp increase in resistivity. Thus we determined $T_s^{\alpha \rightarrow \gamma}$ as the temperature at which the heating curve starts to deviate from a linear relation (dotted line) which is shown in figure 1(b) and (c) under magnetic field of 10 T and without magnetic field, respectively.

Similar measurements have been made by varying the magnetic field, and the transformation start temperatures ($T_s^{\alpha \rightarrow \gamma}$ and $T_s^{\gamma \rightarrow \alpha}$) are plotted by square marks as a function of magnetic field in figure 3. In the figure, ΔT is the difference between the transformation start temperature under various magnetic field strengths $T_s(H)$ and the one measured under zero magnetic field $T_s(H = 0)$. As seen in the figure, ΔT increases with increasing magnetic field as reported previously in some iron based alloys. [4-6]

Among the six panels in figure 3, it should be noted that $T_s^{\alpha \rightarrow \gamma}$ (a) and $T_s^{\gamma \rightarrow \alpha}$ (b) of the Fe-2Rh alloy and $T_s^{\alpha \rightarrow \gamma}$ (c) of the Fe-5Rh alloy are above T_c of each alloy. Thus the α-phase is paramagnetic at these transformation temperatures. In this condition, the transformation temperatures increase parabolically with increasing magnetic field. On the other hand, it should be noted that $T_s^{\gamma \rightarrow \alpha}$ (d) of the Fe-5Rh alloy, $T_s^{\alpha \rightarrow \gamma}$ (e), and $T_s^{\gamma \rightarrow \alpha}$ (f) of the Fe-10Rh alloy are below T_c of each alloy. Thus the α-phase is ferromagnetic at these transformation temperatures. In this case, the transformation temperatures increase linearly with increasing magnetic field. In this way, the field-dependence of transformation temperature for the former case (α-phase is paramagnetic) is obviously different from the latter case (α-phase is ferromagnetic), as in the Fe-Co alloys reported previously [8].

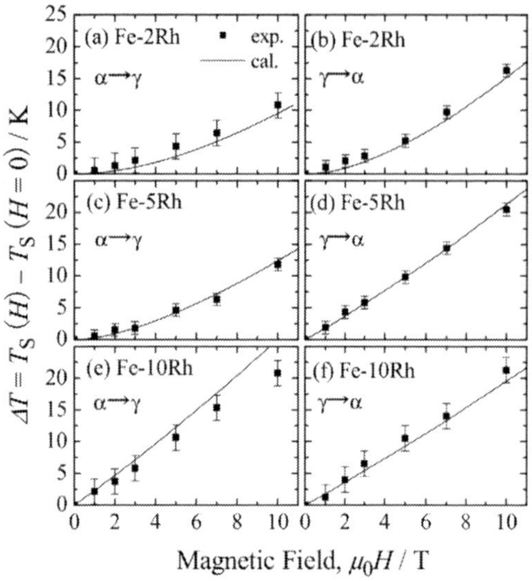

Figure 3. Magnetic field-dependence of transformation temperatures in Fe-2Rh, Fe-5Rh and Fe-10Rh alloys. Marks are obtained from figures 1 and 2. Error bars indicate precision for determining the transformation temperatures from the resistivity curves. Lines are calculated value using equation (2).

2.3. Magnetization and latent heat

Magnetization measurement of the Fe-2Rh, Fe-5Rh, and Fe-10Rh alloys has been made in the heating process near the transformation temperature and at 4.2 K. Typical results of magnetization curves are shown in figure 4(a), (b), and (c) for the Fe-2Rh, Fe-5Rh, and Fe-10Rh, respectively. In the figure, the Curie temperature T_c is also shown. In the case of the Fe-2Rh and Fe-5Rh alloys, all the curves are of the α-phase. On the other hand, in the case of the Fe-10Rh alloy the magnetization curve at 983 K is of the γ-phase while others are of the α-phase. Obviously, the magnetization of the γ-phase is negligibly small compared to that of the α-phase regardless of whether the α-phase is ferromagnetic or paramagnetic. The spontaneous magnetization at 4.2 K, M_0, for each alloy is obtained from the magnetization curve to be 2.25 μ_B/atom for Fe-2Rh, 2.27 μ_B/atom for Fe-5Rh, and 2.33 μ_B/atom for Fe-10Rh alloy. These values lie on the Slater-Pauling curve[9].

Differential scanning calorimetery (DSC) has been made with a heating and cooling rate of 10 K/min by using a NETZSCH-STA449C/3/G Jupiter to obtain the latent heat of the transformation, which is shown in figure 5 (a), (b), and (c) for the Fe-2Rh, Fe-5Rh, and Fe-10Rh alloy, respectively. A clear peak due to the γ→α transformation is seen in the cooling curves and that due to the α→γ one is seen in the heating curves. We notice that the absolute value of latent heat $|L|$ in the cooling process is larger than that in the heating process for all the alloys. The difference is especially large for the Fe-5Rh alloy in which the α-phase is ferromagnetic at $T_s^{\gamma\to\alpha}$ although it is paramagnetic at $T_s^{\alpha\to\gamma}$. The large difference in $|L|$ means that the change in magnetic entropy at $T_s^{\gamma\to\alpha}$ is larger than that at $T_s^{\alpha\to\gamma}$ probably due to magnetic entropy change. Incidentally, the α→γ and γ→α transformation start temperatures and the Curie temperature evaluated from the DSC curves are in good agreement with those obtained from the electrical resistivity measurement.

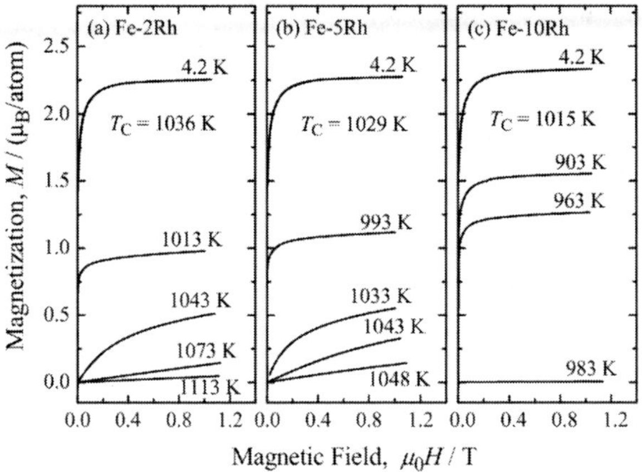

Figure 4. Magnetization curves of (a) Fe-2Rh, (b) Fe-5Rh, (c) Fe-10Rh alloys at various temperatures.

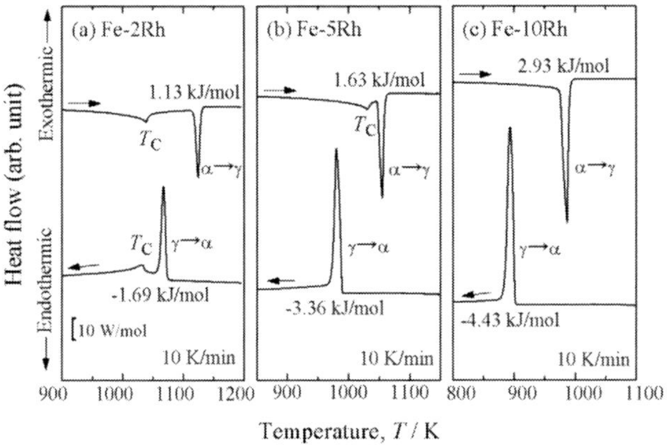

Figure 5. DSC curves of Fe-2Rh, Fe-5Rh and Fe-10Rh alloys

2.4. Thermodynamic analysis

In this section we will discuss the effect of magnetic field on the γ-α transformation temperature of Fe-xRh alloys (x = 2, 5, 10) considering magnetostatic energy. As known from figures 1 and 2, all the alloys examined here exhibit a large temperature hysteresis. Thus we discuss magnetic field-dependence of $T_s^{\alpha\to\gamma}$ and $T_s^{\gamma\to\alpha}$ instead of equilibrium temperature, as in martensitic transformations in ferrous alloys.

The existence of temperature hysteresis means that a driving force is required for the transformation. The driving force is given by the difference in chemical free energy between the two phases $\Delta G(T_s)$ at the transformation start temperature T_s. Let the transformation start temperature under a magnetic field of H be T_s' and we assume that the driving force at T_s' under the magnetic field is equal to $\Delta G(T_s)$. That is, we assume that the driving force does not change by the application of magnetic field. This assumption has been confirmed to be appropriate for martensitic transformation in some iron based alloys[3]. Under the assumption, we can derive the following equation to estimate the field-dependence of transformation temperature, as in martensitic transformations.

$$\Delta G(T_s) - \Delta G(T_s') = -\int_0^H \Delta M dH \qquad (1),$$

where ΔM is the difference in magnetization between γ-and α-phases. The left hand side of the equation can be approximated as $\Delta S\ (T_s' - T_s)$ if the change in transformation temperature is small. Then the field-dependence of transformation temperature is given by

$$\Delta T = T_s' - T_s = -\frac{1}{\Delta S}\int_0^H \Delta M dH \qquad (2).$$

We can calculate field-dependence of transformation temperatures if we know ΔS and ΔM. The change in entropy ΔS has been evaluated as $\Delta S = L/T_s$, where L is the latent heat of γ→α (α→γ) transformation in case we are considering $T_s^{\gamma\to\alpha}$ ($T_s^{\alpha\to\gamma}$).

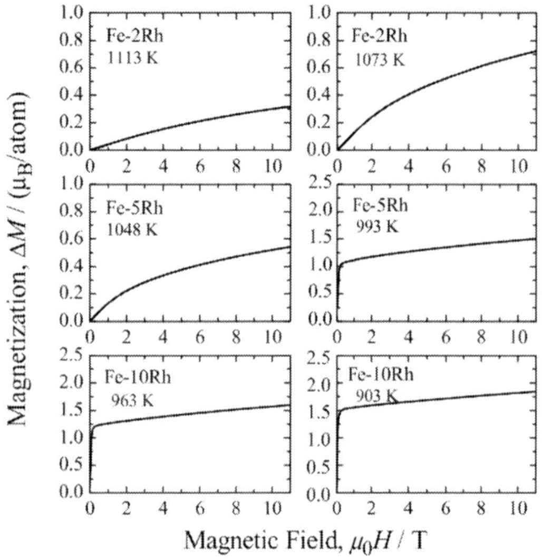

Figure 6. Magnetic field-dependence of the difference in magnetization ΔM between γ- and α-phases of the present alloys calculated from the Arrott plot. The left panels are ΔM's at temperatures near $T_s^{\alpha\to\gamma}$ and the right panels are those at temperatures near $T_s^{\gamma\to\alpha}$.

In obtaining the change in magnetization ΔM, we neglect the magnetization of the γ-phase because it is small compared with the magnetization of the α-phase. The magnetization of the α-phase for $H <$ 1.2 T is obtained by the magnetization curves shown in figure 4. However, that for $H >$ 1.2 T has not been obtained in the present study. To evaluate the value in the high field region, we assumed that the linear relation of M^2 vs H/M (Arrott plot) obtained for $H <$ 1.2 is also satisfied for $H >$ 1.2 T. In this approximation, we can calculate the field-dependence of ΔM at the transformation start temperatures of the present alloys, which is shown in figure 6

Putting the values of ΔS and ΔM thus obtained into equation (2), the field-dependence of the shift in transformation temperature (ΔT) has been calculated for the present alloys, and the results are shown by solid curves in figure 3. The calculated value is in good agreement with experimental result for all the alloys.

3. Rearrangement of martensite variants in Fe-Pd under a magnetic field

A ferromagnetic shape memory alloy of Fe-31.2(at.%)Pd exhibits a thermoelastic martensitic transformation [10] from a cubic phase ($Fm\bar{3}m$) to a tetragonal phase (I4/mmm) at 230 K (T_M) as shown in the figure 7. The tetragonal phase is ferromagnetic with a large uniaxial magnetocrsytalline anisotropy, and its a-axis is the magnetization easy axis.

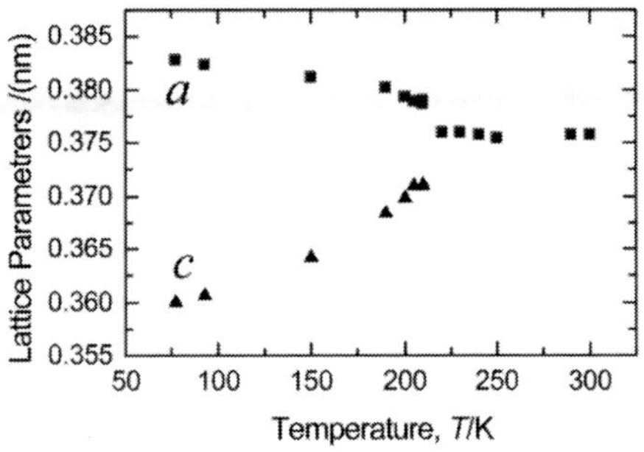

Figure 7. Temperature dependence of lattice parameters in a Fe-31.2Pd (at.%) alloy.

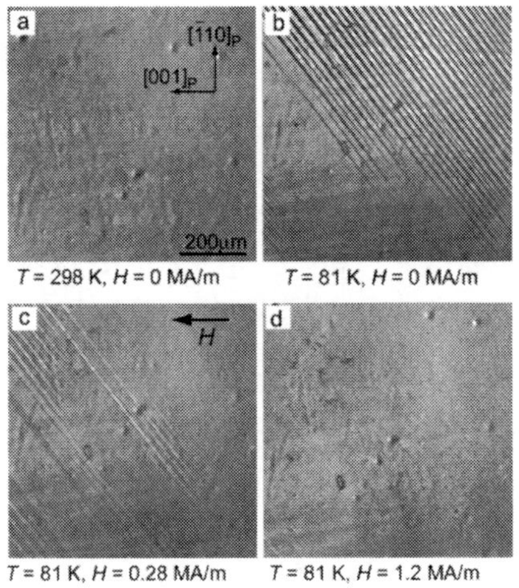

Figure 8. A series of optical micrograph of Fe-31.2Pd alloy showing rearrangement of variants under magnetic field.

The thermally-induced martensite phase usually composed of several orientation variants, which are connected one another by a twinning plane. Figure 8(b) shows such a multi-variant state of the martensite phase of Fe-31.Pd formed from a plane matrix of the parent phase (figure 8(a)) by cooling. The specimen used was a single crystal, which was grown by a floating zone method from an ingot prepared by arc melting. We applied a magnetic field in the $[001]_P$ direction of the parent phase and observed the change in microstructure. This direction corresponds to the magnetization easy axis for one variant and hard axis for another variant. When the magnetic field exceeded about 0.28 MA/m, rearrangement of variants initiates by the movement of twinning planes as reported previously in Ni-Mn-Ga alloys [11-14] and a Fe-30Pd alloy[11]. When the field exceeds about 0.4 MA/m, the whole area comprises the dark region as shown in figure 8(d), meaning that the rearrangement of variants is realized. In association with the rearrangement, the specimen elongated by about 3 % in the field direction, the strain of which was measured by a capacitance method.

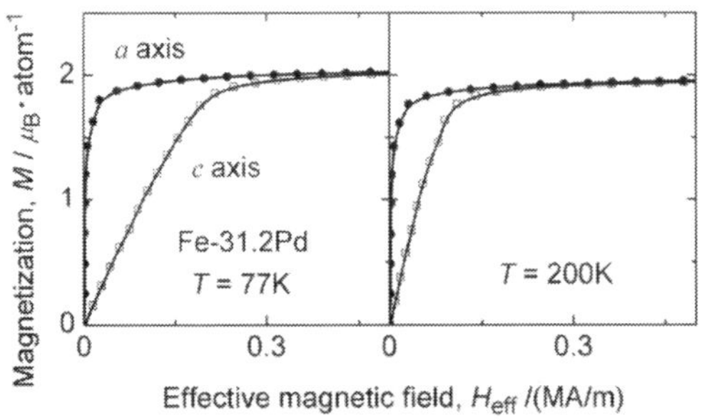

Figure 9. Magnetization curves of Fe-31.2Pd alloy measured in the *a*-axis and *c*-axis at 77 K and 200 K.

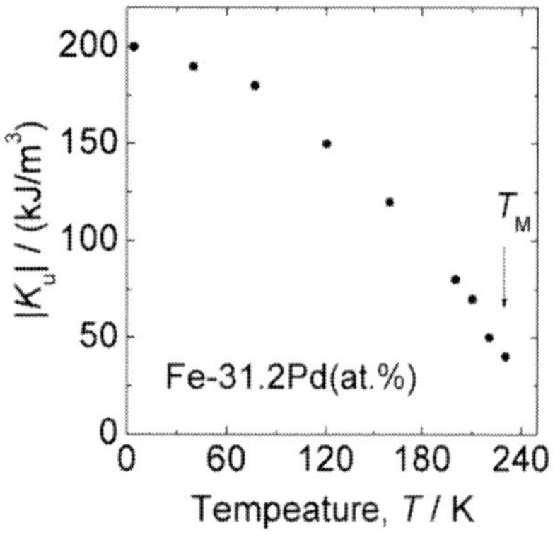

Figure 10. Temperature dependence of uniaxial magnetocrystalline anisotropy constant K_u.

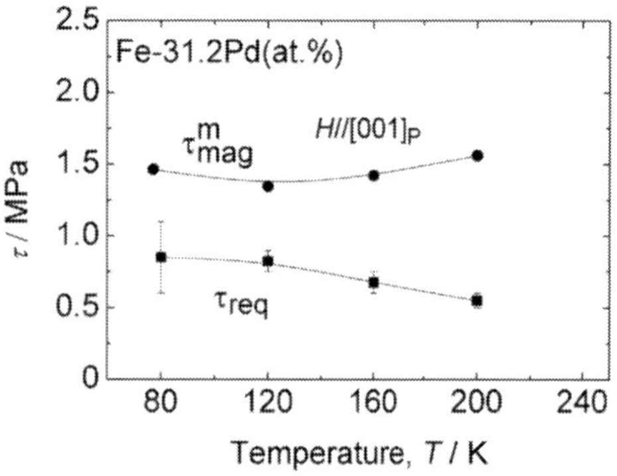

Figure 11. Relation between the maximum of magnetic shear stress $\tau_{mag}{}^{m}$ and the stress required for twinning plane movement τ_{req} in Fe-31.2Pd alloy..

In order to understand the rearrangement of variants by magnetic field, we introduce a magnetic shear stress τ_{mag} acting across the twinning plane. The reason for introducing τ_{mag} is that we usually evaluate a shear stress for analyzing a twinning plane movement, by which the rearrangement of variants proceeds. Using the value of τ_{mag}, the condition for rearrangement of variants by magnetic field will be given as: the value of τ_{mag} is larger than the shear stress required for the twinning plane movement, τ_{req} ($\tau_{mag} > \tau_{req}$). In the following we will show that this condition is certainly satisfied in the present Fe-31.2Pd alloy.

The value of τ_{mag} will be evaluated by $\Delta U_{mag}/s$, where ΔU_{mag} is the magnetic energy difference per unit volume between the two variants separated by the twinning plane considered, and s is the corresponding twinning shear. Assuming that the magnetic energy is mainly composed of magnetocrystalline anisotropy energy and Zeeman energy, the maximum of ΔU_{mag} is equal to the uniaxial magnetocrystalline anisotropy constant $|K_u|$ in the case that a magnetic field is applied to $[001]_P$. The value of s is given by the lattice parameters shown in figure 7 as $s = \{1-(c/a)^2\}/(c/a)$. Then the maximum value of magnetic shear stress, $\tau_{mag}{}^{m}$, will be $|K_u|/s$ under the $[001]_P$ field.

The value of $|K_u|$ was obtained from the area enclosed by the two magnetization curves: along hard magnetization axis and along easy magnetization axis. In order to obtain these curves, we must make magnetization experiments of a single variant state in martensite phase, which was realized by applying a compressive stress along one of $\langle 001 \rangle_P$ directions. Examples of the magnetization curves along the a and c axes are shown in figure 9. By making the same experiment at various temperatures below T_M, we obtained temperature dependence of $|K_u|$ as shown in figure 10. Then, using the value of $|K_u|$ and s, $\tau_{mag}{}^{m}$ is calculated and the result is shown in figure 11.

The other important quantity is the shear stress required for the twinning plane movement. In evaluating this τ_{req}, we have made tensile tests along the $[001]_P$ direction. At any temperature examined, a stage corresponding to the rearrangement of variants appears. From the stress of the stage, we have obtained the stress of τ_{req}, and its temperature dependence is shown in figure 11. Comparing $\tau_{mag}{}^{m}$ and τ_{req} shown in figure 11, it is obvious that the value of $\tau_{mag}{}^{m}$ is larger than τ_{req} at any temperature below T_M. In this way, we have confirmed the propriety of the condition for the occurrence of rearrangement of variants by magnetic field mentioned before regardless of temperature and orientation of magnetic field.

4. Conclusions

We have investigated effects of magnetic field on the $\gamma \rightarrow \alpha$ and $\alpha \rightarrow \gamma$ transformation temperatures in Fe-2Rh, Fe-5Rh and Fe-10Rh alloys by electrical resistivity measurements under magnetic field. As a result, we found that the change in transformation temperature ΔT is almost proportional to magnetic field when the α-phase is ferromagnetic at the transformation temperature. On the other hand, ΔT is proportional to square of magnetic field when the α-phase is paramagnetic. These field-dependencies of transformation temperature are well explained by Clausius-Clapeyron-like equation. We also examined the rearrangement of martensite variants by magnetic field in Fe-31.2Pd alloy and found that magnetic shear stress is larger than the shear stress required for the twinning plane movement in this alloy.

Acknowledgements

The present study is partly supported by the Formation of Worldwide Renowned Centers of Research–The Global COE Program (Project: Center of Excellence for Advanced Structural and Functional Materials Design) from MEXT of Japan.

References

[1] Sadovsky V D, Smirov L V, Fokina Ye A, Malinen P A, and Soroskin I P 1967 *Fiz. Met. Metalloved* **27** 918
[2] Kakeshita T, Kuroiwa K, Shimizu K, Ikeda T, Yamagishi A, Date M 1993 *Mater. Trans. JIM.* **34** 423
[3] Kakeshita T, Saburi T, Kindo K Endo S 1999 *Phase Transitions* **70** 65
[4] Enomoto M, Guo H, Tazuke, Abe Y R, Shimotomai M 2001 *Metall. Mater. Trans.* **32A** 445
[5] Choi J-K, Ohtsuka H, Xu Y, Choo W-Y 2000 *Scripta Mater.* **43** 221
[6] Hao X J, and Ohtsuka H 2004 *Mater. Trans.* **45** 2622
[7] Wijin H P J (Ed.), *Magnetic Properties of Metals, d-Elements, Alloys and Compounds* (Springer-Berlin 1991), p. 12.
[8] Fukuda T, Yuge M, Lee J-H, Terai T, and Kakeshita T, 2006 *ISIJ Int.* **46** 1267
[9] C. Kittel, *Introduction to Solid State Physics*, 7th edition, (Wiley, New York, 1996), p.627
[10] Sugiyama M, Oshima R, Fujita E F 1986 *Trans. Jpn. Inst. Metals* **27** 719.
[11] Ullakko K, Huang J K, Kantner C, O'Handley R C, Kokorin V V 1996 *Appl. Phys. Lett.* **69** 1966
[12] Murray S J, Marioni M A, Kukla A M, Robinson J, O'Handley R C and Allen S M 2000 *J. Appl. Phys.* **87** 5774
[13] Okamoto N, Fukuda T, Kakeshita T 2004 *Sci. Technol. Adv. Mater.* **5** 29
[14] James R D, Wuttig M 1998 *Phil. Mag.* A **77** 1273

Journal of Physics: Conference Series **156** (2009) 012013 doi:10.1088/1742-6596/156/1/012013

Effect of magnetic field on *TTT* diagram of successive $\gamma \rightarrow \varepsilon' \rightarrow \alpha'$ martensitic transformation in SUS304L stainless steel

J H Lee, T Fukuda and T Kakeshita

Department of Materials Science and Engineering, Graduate School of Engineering, Osaka University, 2-1 Yamada-oka, Suita, Osaka 565-0871 Japan

E-mail: ljhcom@mat.eng.osaka-u.ac.jp

Abstract. We have investigated the effect of magnetic field on *TTT* diagram of successive $\gamma \rightarrow \varepsilon' \rightarrow \alpha'$ martensitic transformation in a solution-treated SUS304L stainless steel in order to clarify kinetics of the successive martensitic transformation. The *TTT* diagram of successive $\gamma \rightarrow \varepsilon' \rightarrow \alpha'$ martensitic transformation shows a C-curve with a nose temperature located at 103 K. Incubation time of the successive $\gamma \rightarrow \varepsilon' \rightarrow \alpha'$ martensitic transformation shortens with increasing the strength of magnetic field. However, the nose temperature does not change significantly under the magnetic field. An *in-situ* optical micrograph suggests that the successive $\gamma \rightarrow \varepsilon' \rightarrow \alpha'$ transformation probably proceeds by isothermal $\gamma \rightarrow \varepsilon'$ martensitic transformation followed by athermal $\varepsilon' \rightarrow \alpha'$ martensitic transformation.

1. Introduction

Martensitic transformations are usually classified into two groups from a view point of kinetics: athermal and isothermal ones [1]. The former transformation occurs instantaneously when a definite temperature (M_s) is reached, while the latter occurs after some finite incubation time during isothermal holding. Although the two kinds of transformation kinetics are now basically explained by a universal phenomenological theory [2], the number of the alloys exhibiting an isothermal transformation is small. For this reason, most of the isothermal transformations studied so far are the $\gamma \rightarrow \alpha'$ ones observed in Fe-Ni-Cr and Fe-Ni-Mn alloys. The isothermal transformation in these alloys is characterized by a C-curve in time-temperature-transformation (*TTT*) diagram [3-4].

Recently, however, we found a new type of isothermal transformation in a solution-treated SUS304L stainless steel. In the steel, the successive $\gamma \rightarrow \varepsilon' \rightarrow \alpha'$ martensitic transformation proceeds by isothermal holding in the temperature range between 70 and 170 K [5]. As far as the authors are aware, this was the first finding of successive $\gamma \rightarrow \varepsilon' \rightarrow \alpha'$ martensitic transformation by isothermal holding. However, the *TTT* diagram has not been constructed in previous studies. Also, the kinetics of successive $\gamma \rightarrow \varepsilon' \rightarrow \alpha'$ martensitic transformation has not been clarified yet.

In the present study, therefore, we have constructed the *TTT* diagram of the successive $\gamma \rightarrow \varepsilon' \rightarrow \alpha'$ martensitic transformation in a solution-treated SUS304L and examined the effect of magnetic field on the *TTT* diagram in order to obtain information about kinetics of successive $\gamma \rightarrow \varepsilon' \rightarrow \alpha'$ martensitic transformation.

2. Experimental Procedure

The material used in the study is SUS304L austenitic stainless steel with a composition of 0.023C-0.48Si-1.07Mn-0.005S-8.47Ni-18.20Cr-Bal.Fe (mass %). From the sheet, specimens of 3 × 3 × 1 mm in size were cut out, and were solution-treated at 1323 K for 0.5 h in vacuum followed by quenching into iced water. Then the oxidized surface layer was eliminated by electropolishing in an electrolyte composed of 85% C_2H_5OH and 15% $HClO_4$ in volume. Isothermal holding was carried out under magnetic fields of 0 and 0.8 MA/m in the temperature range between 70 and 170 K. The volume fraction of the α′-martensite, $f_{\alpha'}$, formed by an isothermal holding was evaluated by a magnetization measurement at 300 K (= T_R). That is, $f_{\alpha'} = M_0(T_R) / M_0^{\alpha'}(T_R)$, where $M_0(T_R)$ and $M_0^{\alpha'}(T_R)$ are the spontaneous magnetization of the specimen and that of the α′-phase, respectively, at T_R. Here, the magnetizations of γ-phase and ε′-martensite are neglected because they are nonmagnetic [6]. The value of $M_0^{\alpha'}(T_R)$ is roughly approximated to be the value at 0 K, $M_0^{\alpha'}(0)$, which is estimated to be 1.79 μ_B/atom considering the Slater-Pauling curve and their valence electron concentration [7-8]. Morphology of the martensites was observed by *in-situ* optical microscopy during an isothermal holding process.

3. Results and Discussion

Isothermal holding experiments for constructing the *TTT* curve under magnetic fields have been carried out at several temperatures in the temperature range between 70 and 170 K under magnetic fields of 0 and 0.8 MA/m. Figure 1 shows typical magnetization curves at room temperature (T_R) after isothermal holding in the absence of magnetic field at 103 K for 50 s (a), 170 s (b), 520 s (c), 1660 s (d), 5200 s (e) and 10400 s (f). We can evaluate the volume fraction of α′-martensite ($f_{\alpha'}$) from the spontaneous magnetization as described in experimental procedure. As an example, the spontaneous magnetization obtained from curve (f) is indicated with an arrow in figure 1. Obviously, the spontaneous magnetization increases with increasing isothermal holding time. It means that the $f_{\alpha'}$ increases with increasing the isothermal holding time.

The $f_{\alpha'}$ obtained from the figure 1 is shown with solid circles as a function of holding time in figure 2. The same holding experiments at 103 K have been made under magnetic field of 0.8 MA/m, and the result is also shown with opened circles in figure 2. From the linear relation shown in figure 2, the time required for the formation of 0.5 vol.% of α′-martensite is obtained. It is 6.1 and 3.6 ks at 103 K under magnetic fields of 0 and 0.8 MA/m, respectively, as shown by arrows. We know from the result that the $f_{\alpha'}$ obviously depends on magnetic field as well as holding time.

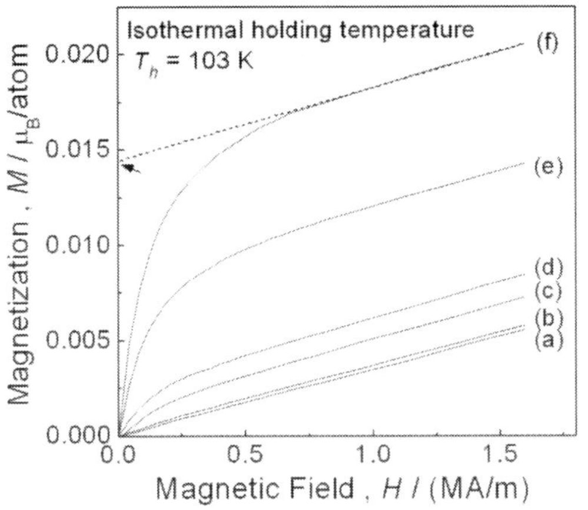

Figure 1. Magnetization curves at 300 K of the solution-treated SUS304L after isothermal holding at 103 K for (a) 50 s, (b) 170 s, (c) 520 s, (d) 1660 s, (e) 5200 s and (f) 10400 s.

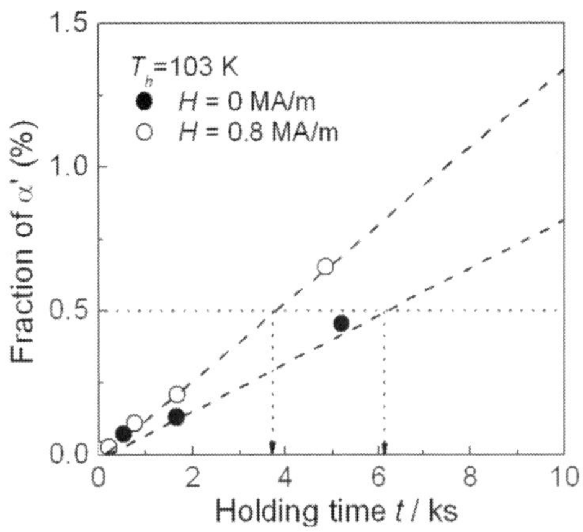

Figure 2. Relation between holing time and volume fraction of α′-martensite during the isothermal holding at 103 K under magnetic fields of 0 and 0.8MA/m.

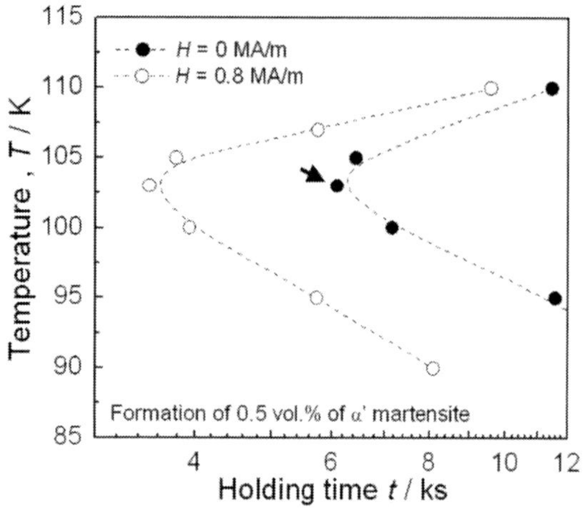

Figure 3. *TTT* diagram of the isothermal martensitic transformation in the solution-treated SUS304L under magnetic fields of 0 and 0.8 MA/m, and dashed lines are guide for eyes.

The same experiments have been made at some temperatures under magnetic fields of 0 and 0.8 MA/m, and the *TTT* diagram of 0.5 vol. % of α'-martensite has been constructed as shown in figure 3. It should be noted in this result that *TTT* diagram of successive $\gamma \rightarrow \varepsilon' \rightarrow \alpha'$ martensitic transformation clearly forms a typical C-curve, and nose temperature (indicated by an arrow) is about 103 K. Moreover, incubation time shortens by the application of magnetic field. However, nose temperature of C-curves dose not change significantly. This result is different from that of Fe-Ni-Cr and Fe-Ni-Mn alloys, whose incubation time shortens and nose temperature decrease with increasing the strength of magnetic field [3-4]. This difference should be ascribed to the difference in kinetics of martensitic transformation between them.

In order to know how the isothermal $\gamma \rightarrow \varepsilon' \rightarrow \alpha'$ martensitic transformation proceeds, we have made *in-situ* optical microscope observation. Figure 4 shows a series of optical micrographs showing isothermal $\gamma \rightarrow \varepsilon' \rightarrow \alpha'$ martensitic transformation at 103 K. After isothermal holding of 3.0×10^2 s, banded plates of the ε'-martensite appear in the γ-phase in figure 4 (a). The number of ε'-plates gradually increases with increasing holding time, and moreover the constant of certain area enclosed by banded ε'-plates (indicated by a dashed rectangle in figure 4 (b)) also increases with increasing the holding time. Such a change in contrast will be due to the gradual growth of banded ε'-martensites. This result suggests that the $\gamma \rightarrow \varepsilon'$ martensitic transformation is isothermal transformation. On the other hand, figure 4 (d) shows that wedge-shaped plates, which are characteristics of α' martensite [9],

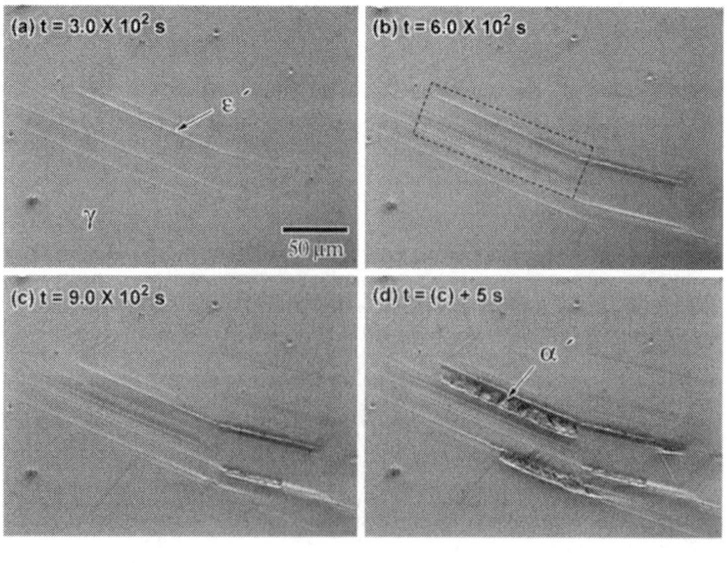

Figure 4. A series of *in-situ* optical micrographs of SUS304L stainless steel showing successive $\gamma \rightarrow \varepsilon' \rightarrow \alpha'$ martensitic transformation at 103 K.

instantaneously formed in the banded ε′-plates. This result implies that the ε′→α′ martensitic transformation is athermal one. We confirmed from figure 4 that the successive γ→ε′→α′ martensitic transformation proceeds by an isothermal γ→ε′ martensitic transformation followed by an athermal ε′→α′ martensitic transformation. Probably, the athermal nature of ε′→α′ martensitic transformation is the reason why the nose temperature dose not change by the application of magnetic field as seen in Figure 3.

4. Conclusions

We have investigated the effect of magnetic field on *TTT* diagram of successive γ→ε′→α′ martensitic transformation in a solution-treated SUS304L stainless and the following results have been obtained.
(i) *TTT* diagram of the successive martensitic transformation shows a C-curve, whose nose temperature is about 103 K.
(ii) Incubation time of the successive γ→ε′→α′ martensitic transformation shortens with increasing the strength of magnetic field. However, the nose temperature does not change significantly under the magnetic field.
(iii) Successive γ→ε′→α′ martensitic transformation probably proceeds by isothermal γ→ε′ martensitic transformation followed by athermal ε′→α′ martensitic transformation.

Acknowledgements
This study was supported by Priority Assistance for the Formation of Worldwide Renowned Centers of Research - The Global COE Program (Project: Center of Excellence for Advanced Structural and Functional Materials Design) from the Ministry of Education, Culture, Sports, Science and Technology (MEXT), Japan.

References

[1] Kajiwara S 1984 *Acta Metall.* **32** 407
[2] Kakeshita T, Kuroiwa K, Shimizu K, Ikeda T, Yamagishi A and Date M 1993 *Mater. Trans. JIM* **34** 423
[3] Kakeshita T, Shimizu K, Ono M and Date M 1992 *Mater. Trans. JIM* **33** 461
[4] Kakeshita T, Sato Y, Saburi T, Shimizu K, Matsuoka Y and Kindo K 1999 *Mater. Trans. JIM* **40** 100
[5] Lee J H, Fukuda T, Kakeshita T and Kindo K 2007 *Mater. Trans. JIM* **48** 2833
[6] Reed R P 1962 *Acta Metall.* **10** 865
[7] Menard J and Weil L 1960 *Adv. Cryog. Eng.* **6** 587
[8] Crangle J and Hallam G C 1963 *Proc. Poy. Soc.* **A272** 119
[9] De A K, Murdock D C, Mataya M C, Speer J G and Matlock D K 2004 *Scr. Mater.* **50** 1445

Journal of Physics: Conference Series **156** (2009) 012014 doi:10.1088/1742-6596/156/1/012014

Biological effects of strong static magnetic fields on insulin-secreting cells

T Sakurai[1,2] **and J Miyakoshi**[1,2]

[1] Department of Radiological Life Sciences, Division of Medical Life Sciences, Graduate School of Health Sciences, Hirosaki University, 66-1 Hon-cho, Hirosaki, 036-8564, Japan

[2] Research Center for Biomedical Sciences, Graduate School of Health Sciences, Hirosaki University, 66-1 Hon-cho, Hirosaki, 036-8564, Japan

E-mail: miyakosh@cc.hirosaki-u.ac.jp

Abstract. The magnetic flux density of MRI for clinical diagnosis has been increasing. However, there remains very little biological data regarding the effect of strong static magnetic fields (SMFs) on human health. To evaluate the biological effects of strong SMFs, we cultured INS-1 cells under exposure to sham and SMF conditions for 1 or 2 h, and analyzed insulin secretion, mRNA expression, cell proliferation and cell number. Exposure to SMF with a high magnetic field gradient for 1 h significantly increased insulin secretion and insulin 1 mRNA expression. Exposure to SMF did not affect cell proliferation and cell number. Our results suggested that MRI systems with a higher magnetic flux density might not cause cell proliferative or functional damages on insulin-secreting cells.

1. Introduction

Magnetic resonance imaging (MRI) has been introduced to many hospitals as a non-invasive diagnostic tool. A strong static magnetic field (SMF) is one of the key components of MRI. Although, most MRI systems used in hospitals at present are made up of a SMF at a magnetic flux density of 1.5 or 3 T in Japan, MRI systems with a much higher magnetic flux density have been under development [1, 2]. During MRI diagnosis, patients are introduced into superconducting magnet and exposed to strong SMF, therefore, it is important to evaluate the biological effects of exposure to a strong SMF.

Various experimental studies to evaluate the effects of SMF have been performed over the last 30-40 years using laboratory animals [3 for review]. In contrast, there have been few studies on the effects of SMF at the cellular level [4 for review]. Meanwhile, information on the biological effects of strong SMF is still lacking.

We have been studying the effects of magnetic fields on insulin-secreting cells [5-8]. Insulin-secreting cells are appropriate for evaluating the effects of SMF on biological systems, because insulin secretion is affected by the behavior of potassium and calcium ions and their channels, which are considered to be a candidate for interaction with SMF [9].

In this study, we cultured INS-1 cells under exposure to sham and SMF conditions, and analyzed insulin secretion, mRNA expression, cell proliferation and cell number to evaluate the effects of exposure to strong SMF on insulin-secreting cells.

2. Materials and Methods

2.1. SMF exposure system

The horizontal exposure system for the strong SMF has been described previously [10]. Briefly, the SMF exposure system consists of a magnet with a bore 150 mm in diameter and 1000 mm long (JMTD-10T150M; Japan Superconductor Technology, Kobe, Japan), a DC power supply (Model 622; Lake Shore Cryotronics, Westerville, OH), cryocooler unit for the magnet (RDK-408D and CSA-71A; Japan Superconductor Technology, Kobe, Japan), a condition control unit (BNA-680S, Espec Engineering, Osaka, Japan), which supplies 5% CO_2 and 95% humidified air for the incubator, and a thermocontroller unit (CL-80R, Taitec, Koshigaya, Japan), which supplies warm water to regulate the temperature of the incubation space. The conditions of magnetic flux density and magnetic field gradient at the exposure positions are summarized in table 1. For the sham-exposure system, we utilized a unit of the same bore shape, but without the magnetic field generator, which is covered by a high-nickel-content soft alloy (Permalloy C, JIS C 2531, Nakano Permalloy, Tokyo, Japan); the SMF in the sham exposure space was less than 5 μT.

Table 1. The magnetic flux density and magnetic field gradient at the center of the exposure positions.

Exposure position	Magnetic flux density (T)	Magnetic field gradient (T/m)	Field-gradient product (T^2/m)
3 T	3	27	81
6 T	6	42	251
10 T	10	0	0

2.2. Insulin secretion

INS-1 cells (donated by Dr. Christopher B. Newgard, Duke University Medical Center) were seeded with 0.5 mL of culture medium on 4-well-MultiDishes (Nalge Nunc No. 176740; Thermo Fisher Scientific Inc., Waltham, MA) at a density of 7.5×10^4 cells/cm^2, and cultured in RPMI 1640 supplemented with 10% fetal bovine serum (FBS), 10 mM 2-[4-(2-hydroxyethyl)-1-piperazinyl]ethanesulfonic acid (HEPES), 1 mM sodium pyruvate, 50 μM 2-mercaptoethanol, 100 U/mL of penicillin, and 100 μg/mL of streptomycin at 37 °C with 95% humidified air and 5% CO_2 for 24 h. After cells were washed once with fresh medium, the culture medium was changed to another 0.5 mL/well of fresh medium. After cells were exposed to SMF for 1 or 2 h, aliquot of culture medium was collected and insulin concentration was measured by enzyme-linked immunosorbent assay (ELISA) kit (Sibayagi, Gunma, Japan). Insulin secretion was standardized to the cell number, which was counted by a cell and particle counter (COULTER Z1; Beckmann Coulter Inc., Fullerton, CA).

2.3. RNA extraction, cDNA synthesis, and PCR reaction

Cells were seeded on 4-well-MultiDishes at a density of 1.5×10^5 cells/cm^2, cultured, and exposed to SMF as described in section 2.2. After exposure to SMF, cells were scraped and total RNA was isolated using an RNeasy® Mini kit (QIAGEN GmbH, Hilden, Germany) according to the manufacturer's instructions. cDNA synthesis was performed using random hexamers as primers and ExScript™ RTase (TaKaRa Bio, Siga, Japan) according to the manufacturer's instructions. Semi-quantitative PCR was performed using a SYBER® Premix Ex Taq™ kit (TaKaRa Bio) and a Smart Cycler® II System (Cepheid, Sunnyvale, CA) according to the manufacturers' instructions. The cycle conditions were as follows; after holding at 95 °C for 10 s, 45 cycles of denaturation at 95 °C for 5 s and primer annealing at 60 °C for 20 s. The primers used in this study are summarized in Table 2. β-Actin and glyceraldehyde-3-phosphate dehydrogenase (GAPDH) expression were used to normalize the amount of template cDNA in the semi-quantitative PCR reaction.

Table 2. Sequences of primers used in real-time reverse transcription polymerase chain reaction.

Gene		Sequence	Gene bank accession
insulin 1	Forward	5'-AAGCGTGGCATTGTGGATCA-3'	NM_019129
	Reverse	5'-GGAGTGGTGGACTCAGTTGCAGTA-3'	
insulin 2	Forward	5'-TCAAACAGCACCTTTGTGGTTCTC-3'	NM_019130
	Reverse	5'-ACCTCCAGTGCCAAGGTCTGA-3'	
β-actin	Forward	5'-TGACAGGATGCAGAAGGAGA-3'	NM_007393
	Reverse	5'-GCTGGAAGGTGGACAGTGAG-3'	
GAPDH	Forward	5'-GGCACAGTCAAGGCTGAGAATG-3'	NM_017008
	Reverse	5'-ATGGTGGTGAAGACGCCAGTA-3'	

2.4. WST-1 cell proliferation assay and Cell number

Cells were seeded, cultured, and exposed to SMF as described in section 2.2. After SMF exposure, 50 μL of premixed WST-1 (TaKaRa Bio) was added to each well. All test plates were incubated for an additional 4 h in a conventional incubator. The optical density was then measured using a microplate reader (Sunrise™ Basic; Tecan, Zurich, Switzerland), at 450 nm. Cell number analysis was performed as described in section 2.2 excluding the measurement of insulin concentration in the culture medium.

2.5. Statistical analysis

Statistical comparisons were performed by analysis of variance and, when appropriate, using the Bonferroni correction method. All results are presented as the mean ± SE, and studies were repeated in at least three independent experiments.

3. Results

3.1. Insulin secretion

Figure 1 represents the insulin secretion during exposures to SMF for 1 or 2 h at 3, 6, or 10 T position. Exposure to SMF for 1 h at the 6 T position increased insulin secretion significantly. At the 3 T position, insulin secretion was slightly, but not significantly, increased by exposure for 1 or 2 h. At the 10 T position, exposure to SMF did not affect insulin secretion.

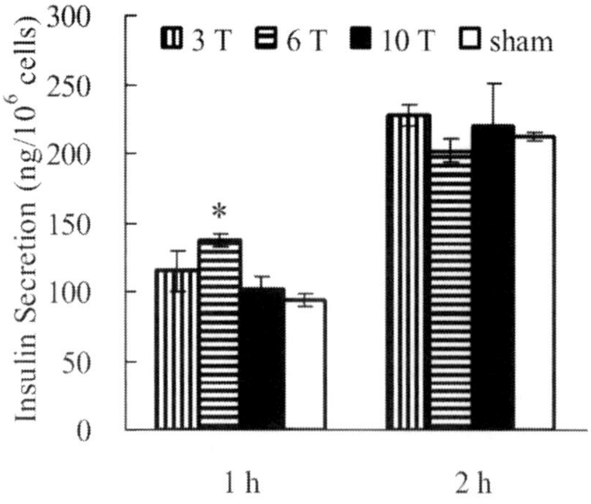

Figure 1. Alterations of insulin secretion levels during exposure to static magnetic field (SMF). INS-1 cells were exposed to SMF at the 3, 6, or 10 T position for 1 or 2 h. Data are represented as the mean ± SE from three independent experiments. * $P < 0.05$.

3.2. Expression of insulin mRNA

We performed semi-quantitative RT-PCR using β-actin and GAPDH as house keeping genes. Only the data standardized to β-actin are shown in figure 2; no differences were observed between the results standardized to β-actin vs. GAPDH (data not shown). Similar to insulin secretion, a significant increase in insulin 1 (ins1) mRNA expression was detected after SMF exposure for 1 h at the 6 T position. A slight but not significant increase in insulin 2 (ins2) mRNA expression was observed after SMF exposure for 1 h at the 6 T position.

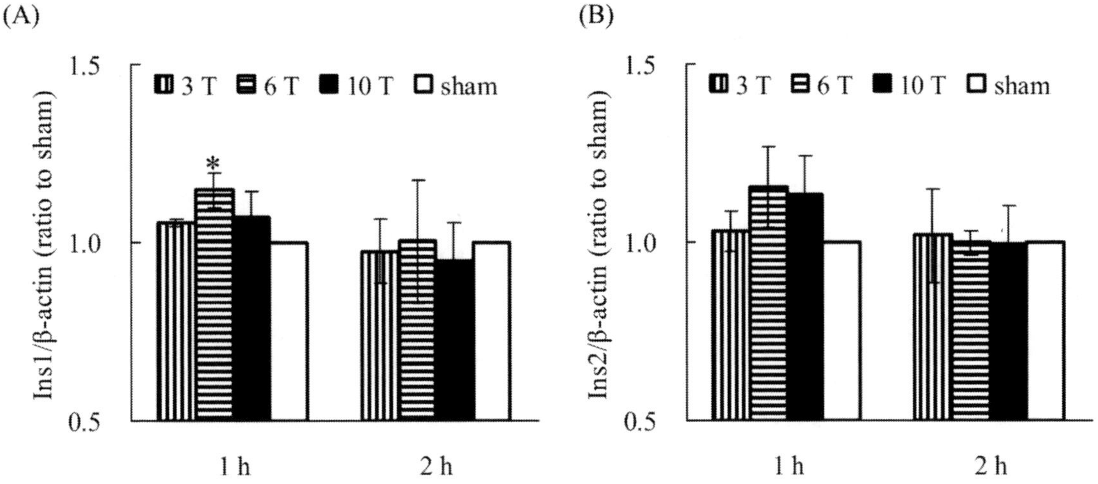

Figure 2. Alterations of the expression of insulin 1 (ins1, A) and insulin 2 (ins2, B) following exposure to static magnetic field (SMF). INS-1 cells were exposed to SMF at the 3, 6, or 10 T position for 1 or 2 h. Insulin mRNA levels were standardized to β-actin and glyceraldehyde-3-phosphate dehydrogenase, and only the results of β-actin are shown. Data are expressed relative to sham, and represented as the mean ± SE from three independent experiments. * $P < 0.05$.

3.3. WST-1 cell proliferation assay and cell number

Insulin-secreting cells store insulin for rapid release after glucose stimulation. Therefore, insulin is released to culture medium when insulin-secreting cells die. If it occurs during exposure to SMFs, observed increase in insulin secretion (figure 1) might be due to death of insulin-secreting cells. To estimate this possibility, we assessed the cell viability by WST-1 cell proliferation and cell number (table 3). Since the WST-1 cell proliferation assay only evaluated the cellular mitochondrial activity, we performed both analyses. No effects of exposure to SMF were observed under any exposure condition, using both the WST-1 cell proliferation assay and cell number. These results suggested that the increase in insulin secretion during exposure to SMFs were not due to insulin-secreting cell death.

Table 3. The effects of static magnetic field (SMF) on cell proliferation and cell number.

Exposure position	Cell proliferation (% of sham exposure)		Cell number ($\times 10^5$ cells/cm^2)	
	1 h	2 h	1 h	2 h
3 T	104.5 ± 4.7	100.4 ± 1.4	3.28 ± 0.06	3.28 ± 0.07
6T	103.8 ± 2.3	101.0 ± 1.1	3.10 ± 0.04	3.15 ± 0.10
10 T	101.4 ± 1.2	99.3 ± 1.8	3.23 ± 0.03	3.21 ± 0.12
Sham	(100)	(100)	3.25 ± 0.05	3.28 ± 0.04

4. Discussion

In this experiment, we evaluated the effect of a strong SMF on insulin-secreting cells by measuring insulin secretion, mRNA expression and cell viability. Exposure to SMF at 6 T position for 1 h significantly increased insulin secretion and ins1 mRNA expression (figures 1 and 2).

We have previously reported that magnetic field gradient and/or the product of magnetic field gradient and magnetic flux density played a critical role in the effects on biological systems [11, 12]. In this study, we also detected the effects of magnetic field gradient and/or the product of magnetic field gradient and magnetic flux density, not magnetic flux density, on insulin secretion. At the 6 T position, the magnetic field gradient and the product of magnetic field gradient and magnetic flux density was the highest, and significant effects on insulin secretion were observed at this position. At the 3 T position, where the magnetic field gradient and magnetic force was modest, a slight increase in insulin secretion was observed; indicating the relationship between magnetic field gradient and/or the product of magnetic field gradient and magnetic flux density and the increase in insulin secretion. The magnetic field gradient and/or the product of magnetic flux density and magnetic field gradient affected the diamagnetic components of cells [13] and the distribution of dissolved oxygen in the culture medium [14]. These phenomena are thought to be possible mechanisms of the effects of SMF.

Regarding the effects on insulin secretion, ins1 mRNA expression was increased by exposure to SMF at the 6 T position for 1 h. The transcription of insulin mRNA has been reported to be regulated by increases in intracellular calcium concentrations, and by secreted insulin itself [15, 16]. Therefore, exposure to SMF might increase intracellular calcium concentration, which, in turn, increases insulin secretion and insulin mRNA, or exposure to SMF might directly increase insulin secretion resulting in increased insulin mRNA expression.

Evaluation of metabolic activity of insulin-secreting cells is possible by combination of WST-1 cell proliferation assay and cell count [17]. From our results of the WST-1 cell proliferation assay and cell count (table 3), exposure to SMF did not affect the metabolic activities of INS-1 cells. Although a reduction of metabolic activity in the human leukemic cell line HL-60 following 72 h exposure to SMF at 1 T has been reported [18], the time period of exposure and the cell line used in the current study were different from this report. This could explain why we did not detect the effects of SMF on metabolic activities in this study.

In conclusion, MRI systems with higher magnetic flux density are currently under development, despite the fact that the effects of exposure to strong SMF on biological systems have not been well clarified. Our results demonstrated that exposure to magnetic fields with high magnetic flux density and magnetic field gradient did not cause any damage to insulin-secreting cells, and suggest that MRI systems with higher magnetic flux density might not cause cell proliferative or functional damages on cells.

Acknowledgments

This study was supported in part by a Grant-in-Aid from the Magnetic Health Science Foundation and a Grant for Hirosaki University Institutional Research.

References

[1] Fu R, Brey W W, Shetty K, Gor'kov P, Saha S, Long J R, Grant S C, Chekmenev E Y, Hu J, Gan Z, Sharma M, Zhang F, Logan T M, Brüschweller R, Edison A, Blue A, Dixon I R, Markiewicz W D and Cross T A 2005 *J. Magn. Reson.* **177** 1
[2] Vaughan T, DelaBarre L, Snyder C, Tian J, Akgun C, Shrivastava D, Liu W, Olson C, Adriany G, Strupp J, Andersen P, Gopinath A, van de Moortele P-F, Garwood M and Ugurbil K 2006 *Magn. Reson. Med.* **56** 1274
[3] Saunders R 2005 *Prog. Biophys. Mol. Biol.* **87** 225
[4] Miyakoshi J 2005 *Prog. Biophys. Mol. Biol.* **87** 213
[5] Sakurai T, Satake A, Sumi S, Inoue K and Miyakoshi J 2004 *Bioelectromagnetics* **25** 160
[6] Sakurai T, Koyama S, Komatsubara Y, Wang J and Miyakoshi J 2005 *Europ. Biol.*

Bioelectromagnetics **1** 60
[7] Sakurai T, Koyama S, Komatsubara Y, Wang J and Miyakoshi J 2005 *Biochem. Biophys. Res. Commun.* **332** 28
[8] Sakurai T, Yoshimoto M, Koyama S and Miyakoshi J 2008 *Bioelectromagnetics* 29 118
[9] WHO. 2006. Environmental Health Criteria 232: STATIC FIELDS, p 2. Available from: http://www.who.int/peh-emf/publications/EHC_232_Static_Fields_full_document.pdf
[10] Nakahara T, Yaguchi H, Yoshida M and Miyakoshi J 2002 *Radiology* **224** 817
[11] Hirose H, Nakahara T, Zang QM, Yonei S and Miyakoshi J 2003 *In vitro Cell. Dev. Biol. - Animal* **39** 348
[12] Sakurai T, Terashima S and Miyakoshi J 2008 *Bioelectromagnetics* **29** 277
[13] Iwasaka M and Ueno S 2005 *J. Appl. Phys.* **97** 10Q907
[14] Ueno S, Iwasaka M and Kitajima T 1994 *J. Appl. Phys.* **75** 7174
[15] Leibiger I B, Leibiger B, Moede T and Berggren P-O 1998 *Mol. Cell.* **1** 933
[16] Leibiger B, Moede S, Uhles S, Berggren P-O and Leibiger I B 2002 *Biochemical. Soc. Transactions.* **30** 312
[17] Segu V B G, Li G and Metz S A 1998 *Metabolism* **47** 824
[18] Sabo J, Mirossay L, Horovcak L, Sarrissky M, Mirossay A and Mojzis J 2002 *Bioelectrochemistry* **56** 227

Effects of Static Magnetic Field on Mutagenesis in *in vitro*

M Ikehata[1], S Yoshie[1], N Hirota[2] and T Hayakawa[1]

[1]Railway Technical Research Institute, Kokubunji, Tokyo 185-8540, Japan

[2]Nano Materials Center, National Institute for Materials Science, Tsukuba, Ibaraki 277-8561, Japan

E-mail: ikehata@rtri.or.jp

Abstract. Effects of static magnetic field up to 13 T were estimated in *Escherichia coli* and *Saccharomyces cerevisiae*. We observed that exposure to a 5 T static magnetic field resulted in a slight but significant increase in gene recombination frequency while frequency of reverse point mutation was not altered in *S. cerevisiae*. This mutagenic effect showed a dose response relationship as J-shape. To investigate an involvement of reactive oxygen species in possible mutagenic effect of static magnetic field, SOD deficient *E. coli* QC774 was used in thymine synthesis-based mutation assay. The result shows that static magnetic field up to 13 T did not indicate mutagenicity. Thus, it is suggested that frequency of point mutation does not changed under static magnetic field regardless of its susceptibility to superoxide. These results suggest that strong static magnetic field would have small but detectable mutagenic potential. Although mechanism of the mutagenic effect of static magnetic field has not been elucidated yet, the extent of this effect is estimated extremely small by comparison with other mutagens such as ultraviolet irradiation.

1. Introduction

The chance for exposure to various magnetic and electromagnetic fields in general public and working environment are increasing. In recent years, the biological effects by exposure to magnetic fields, especially in power frequency and radio frequency were extensively investigated along with growing of social concern. However, there are not enough studies to evaluate the biological effects of static magnetic field while opportunities for exposure to strong magnetic field such as MRI (up to 3 Tesla at present) are increasing. It was reported that exposure to strong static magnetic field caused small increase of somatic recombination in wing spot test of *Drosophila melanogaster* [1], increase mutation frequency in forward mutation assay in SOD (superoxide dismutase) deficient *Escherichia coli* [2] and enhancement of mutagenicity of DNA reactive mutagen in reverse mutation assay in *E. coli* [3]. However, the mechanism of the mutagenic effect has not been determined yet. On the other hand, no mutagenic effect was observed in bacterial mutation assay (Ames' test) using *Salmonella typhimurium* and *E. coli* [3]. Therefore, additional research evidences will be necessary to understand the mutagenic effects of magnetic fields.

In this study, mutagenicity of strong static magnetic fields were investigated using budding yeast *Saccharomyces cerevisiae* as a eukaryote model and SOD-deficient *Escherichia coli* as a ROS (reactive oxygen species) sensitive model organism.

Journal of Physics: Conference Series **156** (2009) 012015 doi:10.1088/1742-6596/156/1/012015

2. Materials and Methods

2.1 Exposure systems

Two types of superconducting magnet were used for exposure to static magnetic fields. One of the magnets is a superconducting magnet (JS-500 (Toshiba)) and generates homogeneous up to 5 T horizontal magnetic field with 5% fluctuation within 10cm from the centre of the bore (Magnet 1; Fig. 1 (a)). This magnet was placed in a temperature-controlled room and temperature of the exposure space was maintained at 30 or 37±1°C. The other magnet is also a superconducting magnet (JMTD-10C13E-NC (JASTEC)) and generates up to 13 T of vertical magnetic field (Magnet 2; Fig. 1 (b)). Temperature in bore was kept at 37°C by running hot water in silicon tube, which covered around internal surface of the bore of the Magnet 2.

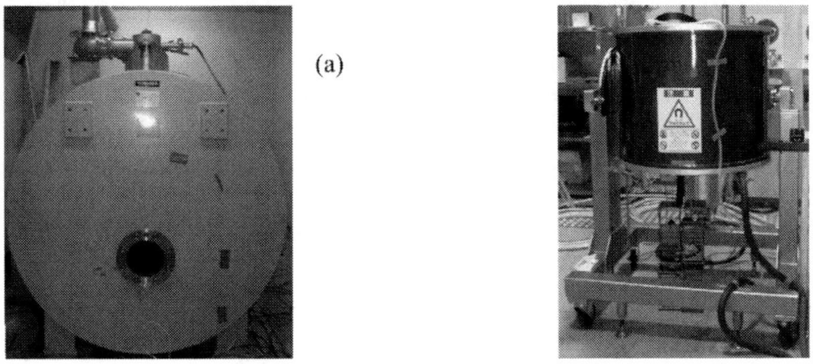

Figure 1. Exposure systems of static magnetic fields; (a) Magnet 1 (5 T), (b) Magnet 2 (5, 10, 13 T).

2.2 S. cerevisiae mutation assay

S. cerevisiae XD83 (*MATa/MATα, lys1-1/lys1-1, arg4-4/arg4-17, RAD*) was obtained from American Type Culture Collection (Virginia, USA) and was used as the tester strain in this study. For magnetic field exposure, the Magnet 1 was used in this experiment. *S. cerevisiae* XD83 was pre-cultured to late log phase (approximately 10^8cells/ml) in YPD medium.

In mutation assay, 0.1 ml of cell suspension was mixed with molten soft agar (0.6 % Bacto-agar, 0.5 % NaCl) and poured on to low lysine synthetic complete plate for detecting point mutation frequency on *lys 1-1*. 0.1 ml of 1/100 diluted cell suspension was mixed with molten soft agar and poured on to low arginine synthetic complete plate for detecting gene conversion frequency on *ARG4* allele (between *arg 4-4* and *arg 4-17*).

At least 6 plates were made for each test condition and randomly divided into two groups. One group was exposed to a static magnetic field for 4 days at 30 ± 1 °C as exposed group. The other group was incubated in a conventional incubator as unexposed control. Number of colonies on each plate was scored as revertant and the mutation frequency was calculated.

2.1. E. coli forward mutation assay

Two strains, *E. coli* QC774 cells (Φ(*sodA-lacZ*)49 Φ(*sodB-kan*)1-Δ_2 Cm^r Km^r) [4], which is defective in SOD gene, *sodA* and *sodB*, and its wild-type (GC4468) were used in this experiment. The SOD is an enzyme which eliminates superoxide from the cells by generation of hydrogen peroxide and oxygen from superoxide and hydrogen ion. The superoxide ion is a kind of the free radicals and causes damage to cellular materials such as DNA.

For magnetic field exposure, the Magnet 2 was used in this experiment. These cells were pre-incubated for 4 hr in Luria-Bertani (LB) medium and the pre-incubated culture was added to fresh LB medium. The plastic tube with suspended culture and fresh LB medium was put into the bore of superconducting magnet for exposure to 0, 5, 10 or 13 T. After 24 hr exposure, cells were poured onto glucose minimum medium plates containing trimethoprim and thymine to select Thy^- (thymine synthesis deficient) mutant and onto LB medium plate after dilution properly for determining survival cell concentration [5]. Mutation frequency was calculated as Thy$^-$ cells count per 10^7 survival cells.

3. Results and Discussion

When *S. cerevisiae* XD83 cells were exposed to a 5 T static magnetic field, reverse mutation frequency in *ARG4* locus slightly increased (approx. 1.4 times higher than control) while reverse point mutation frequency on *lys1-1* was not altered. This result suggests that magnetic field exposure caused chromosomal recombination or gene conversion but not point mutation. This mutagenic effect disappeared on exposure to a 2 T static magnetic field and was even lower than the control on exposure to a 1 T static magnetic field. When cells were exposed to a 0.5 T static magnetic field, there was no difference in the frequency of reverse mutation in both *ARG4* and *lys1*. These results suggest that exposure to a strong static magnetic field shows weak mutagenicity and its threshold would be above 2 T in this tester strain. In addition, the dose response relationship between magnetic field density and mutagenic effect showed J-shape (Fig. 2). In our previous study, extent of mutagenic effect of static magnetic field was increased linearly up to 2 T and saturated in wing spot test of *Drosophila melanogaster* [1] and co-mutagenicity of Ames test [3]. In case of these assays, tester strains have partial deficiency of DNA repair ability and therefore these strains are sensitive to various mutagens. On the other hand, *S. cerevisiae* strain used in this study is proficient for DNA repair and its J-shape dose response relationship between magnetic field density and mutagenic effect was different from DNA repair deficient *Drosophila melanogaster* and *E. coli*. These different responses would be reasonable if the exposure to strong static magnetic field caused an increase of DNA lesion frequency. Obviously, although a strong magnetic field even at 5T does not have enough energy to modify the covalent bond of DNA directly, indirect effects such as increase of oxidative damage by exposure to a strong static magnetic field that was reported by Watanabe et al [6] would relate to the cause of increase in the mutation frequency.

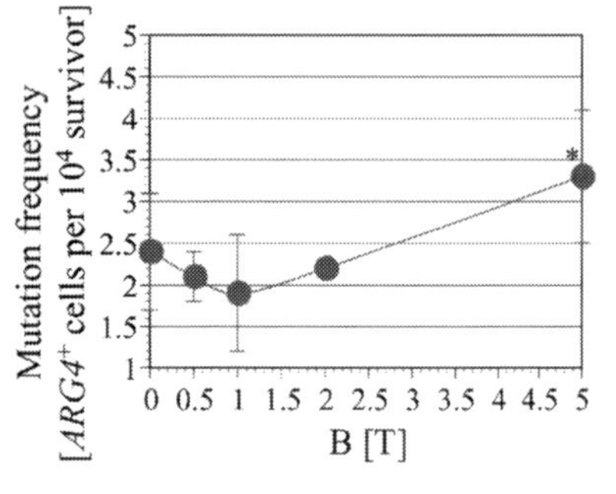

Figure 2. Dose dependent relationship between magnetic flux density and mutation frequency on *ARG4* locus in *S. cerecisiae* XD83.
*; P<0.05 by Student's test

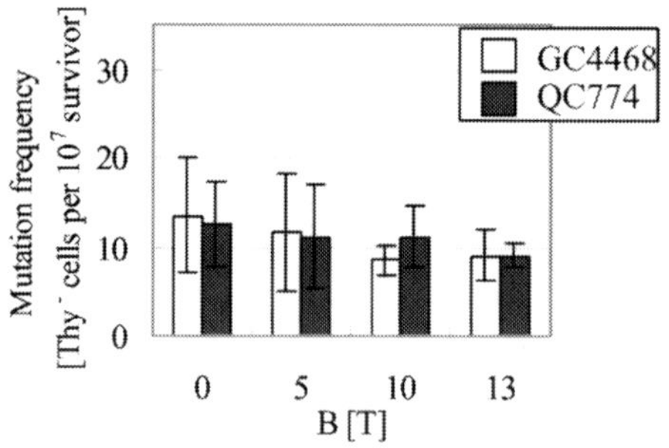

Figure 3. Mutation frequency on thy metabolism pathway in *E. coli* tester strains with/without MF exposure.

On the other hand, no statistically significant difference on *thy* forward mutation was observed between MF exposed and unexposed control cells in 5, 10 and 13 T, respectively in both *E. coli* tester strains (Fig. 3). This suggests that SOD pathway could not relate to the mutagenic effect of magnetic fields. Moreover, these results suggest that mutagenesis on chromosomal level such as recombination and gene conversion would be more sensitive to MFs than mutagenesis of point mutation with lack of SOD pathway. In this study, although we observed the recombinogenic effect of MFs above 2T in *S. cerevisiae*, the degree of the effects was extremely small. Because, in same experiment, weak UV light (ca. 1/10 of sunlight) induced both recombination and point mutation approximately 10 times higher than control in *S. cerevisiae*. Thus, our data suggests that mutagenicity of static magnetic field would be negligibly small to assess environmental risks.

4. Conclusion

Effects of static magnetic field up to 13 T were estimated in *Escherichia coli* and *Saccharomyces cerevisiae*. For point mutation, we observed no effect for mutation frequency by exposure to up to a 5 T static magnetic field in *S. cerevisiae* and up to a 13 T static magnetic field in SOD-deficient *E. coli*. On the other hand, frequency of recombinogenic mutation was slightly changed by static magnetic field up to 5 T depends on its field density.

These results suggest that strong static magnetic field would have small but detectable mutagenic potential; however, the extent of this effect is estimated extremely small by comparison with other mutagens such as ultraviolet irradiation.

References

[1] Takashima Y, Miyakoshi J, Ikehata M, Iwasaka M, Ueno S and Koana T, 2004 *J. Radiat. Res.*, **45** 393
[2] Zhang QM, Tokiwa M, Doi T, Nakahara T, Chang PW, Nakamura N, Hori M, Miyakoshi J and Yonei S, 2003 *Int. J. Radiat. Biol.*, **79** 281
[3] Ikehata M, Suzuki Y, Shimizu H, Koana T and Nakagawa M, 1999 *Mut. Res.*, **427** 147
[4] Carlioz A and Touati D, 1986 *EMBO J.* **5** 623
[5] Miller J, 1972 *Experiments in molecular genetic* (Cold Spring Harbor Laboratory, Cold Spring Harbor, NY) p218
[6] Watanabe Y, Nakagawa M and Miyakoshi Y, 1997 *Industrial Health*, **35** 285

Journal of Physics: Conference Series **156** (2009) 012016 doi:10.1088/1742-6596/156/1/012016

Yeast cells proliferation on various strong static magnetic fields and temperatures

E S Otabe[1], S Kuroki[1], J Nikawa[1], Y Matsumoto[2], T Ooba[3], K Kiso[4] and H Hayashi[5]

[1] Faculty of Computer Science and Systems Engineering, Kyushu Institute of Technology, 680–4 Kawazu Iizuka Fukuoka 820–8502, Japan
[2] Fukuoka University, 8–19–1 Nanakuma, Jonan-ku, Fukuoka 814–0180, Japan
[3] Fukuoka Industrial Technology Center, 1465–5 Aikawa-machi, Kurume, Fukuoka 839–0861, Japan
[4] Fukuoka Regional Taxation Bureau, 2–11–1 Hakataekihigashi, Hakata-ku Fukuoka, 812–8547, Japan
[5] Kyushu Power Electric, 2–1–47 Shiobaru Minami-ku Fukuoka 815-8520, Japan

E-mail: `otabe@cse.kyutech.ac.jp`

Abstract. The effect of strong magnetic fields on activities of yeast cells were investigated. Experimental yeast cells were cultured in 5 ml of YPD(Yeast extract Peptone Dextrose) for the number density of yeast cells of $5.0 \pm 0.2 \times 10^6$/ml with various temperatures and magnetic fields up to 10 T. Since the yeast cells were placed in the center of the superconducting magnet, the effect of magnetic force due to the diamagnetism and magnetic gradient was negligibly small. The yeast suspension was opened to air and cultured in shaking condition. The number of yeast cells in the yeast suspension was counted by a counting plate with an optical microscope, and the time dependence of the number density of yeast cells was measured. The time dependence of the number density of yeast cells, ρ, of initial part is analyzed in terms of Malthus equation as given by $\rho = \rho_0 \exp(kt)$, where k is the growth coefficient. It is found that, the growth coefficient under the magnetic field is suppressed compared with the control. The growth coefficient decreasing as increasing magnetic field and is saturated at about 5 T. On the other hand, it is found that the suppression of growth of yeast cells by the magnetic field is diminished at high temperatures.

1. Introduction

The relation between magnetic field and living organism has been investigated by many groups. It was reported that low-frequency magnetic fields such as 50 Hz and 10 mT affected yeast *Saccharomyces cerevisiae* activities: the the number of yeast cells were decreasing and their growths were restricted[1]. Many reports are cited in Ref. [1] and the effect of the environment magnetic field is discussed.

On the other hand, no effect on the growth of yeast *Saccharomyces cerevisiae* was reported in the AC magnetic field of 50 Hz, 2.45 mT[2]. The effect of AC magnetic field is difficult to detect, since the joule heat is always occurred in measurements and it causes the increase of the temperature.

Recent development of cryocooler allows us to use superconducting technology easier, and strong magnetic field over 10 T in room temperature space can be applied by superconducting magnet[3]. This new technology accelerated many applications of such as physics, chemistry and biology fields. For example, the growth of *Escherichia coli* in magnetic field of 5.2–6.1 T is reported and the death is suppressed 100,000 times than that under a geomagnetic field[4]. This result suggests that the growth rate of *Escherichia coli* is suppressed by magnetic field. In addition, it is reported that strong static magnetic fields (5 T or 9 T) induce mutations through elevated production of intracellular superoxide radicals in *Escherichia Coli*[5].

In the case of effect of static magnetic field on yeast cell activities, Ikehata *et al.* found that the sedimentation pattern of yeast cells at 5 T was changed although the static magnetic field did not affect the expression level of any gene in yeast cells. Iwasaka *et al.* performed the proliferation of yeast cells under the gradient magnetic field at 9–14 T. They found the proliferation of the yeast cells was suppressed and they tried to explain the observed phenomenon by magnetic force by the diamagnetism of oxygen and the gradient magnetic field.

It is desire to investigate the pure effect of the magnetic field to yeast cell without that of the induced environmental changes by the magnetic field. In the present study, yeast cells were proliferated in strong static magnetic field up to 10 T at various temperatures with small effect of the magnetic-induced force. The effect of the static magnetic field is reported and the possible effects of magnetic field are discussed.

2. Experimental

2.1. Japanese Sake mash

First, we started to study what sort of difference in taste would arise when a magnetic field was applied to yeast cells while brewing Japanese sake. Since sake is an extremely delicate brew, there is the potential for major differences to arise. Furthermore, sensory evaluation (sake tasting) by the human is very sensitive and thus able to detect even subtle differences.

Since the sake manufacturing process is rather complex, some Koji (aspergillus oryzae, the mold used in the brewing process) extract was prepared and then some sake mash was provided. Koji extract is made by rice and malted rice that has been mixed together and saccharified at 55°C for 4 days. Koji extract has an amber color and is sweet tasting. The concentration of the glucose was 15%. Then sake yeast K-7 was put for making refined sake directly into it. The number density of the yeast cells was $2.5 \pm 0.1 \times 10^6$/ml. Two 100 ml bottles of sake mash were prepared. The bottles were sealed. Yeast cell converts glucose into alcohol through fermentation. The temperature of the Japanese sake mash was maintained at 15°C and a magnetic field of 10 T was applied. The other controlled sake mash was also prepared without a magnetic field for comparison. Since the cultivation is stationary and the low temperature is kept, the cultivation needs about one week.

The static magnetic field was applied by a superconducting magnet cooled by cryocooler manufactured by Japan Superconductor Technology Inc.(JASTEC). Maximum magnetic field is 10 T. The induced magnetic field is uniform within 1.4% in a sphere with radius of 40 mm. Temperature control of the yeast suspensions is very important factor in the present experiment. Therefore, two jacketed vessels were prepared. One vessel was put into the room temperature bore of the superconducting magnet system, and the other vessel was for control. The jacket water were circulated to two vessels in series and the temperature of the water was controlled by a chiller unit. The fluctuation of the temperature was ±0.1°C. The sake mash bottle was put into the water in the vessel.

The taste test was performed by alcoholic beverage appraisers from the Fukuoka Regional Taxation Bureau. In addition, low-boiling flavor components and organic acid components were measured. The number density of the yeast cells ρ in the sake mash for magnetic field ρ_M and for control ρ_C were also counted by the standard plate count technique with an optical microscope and a counting plate.

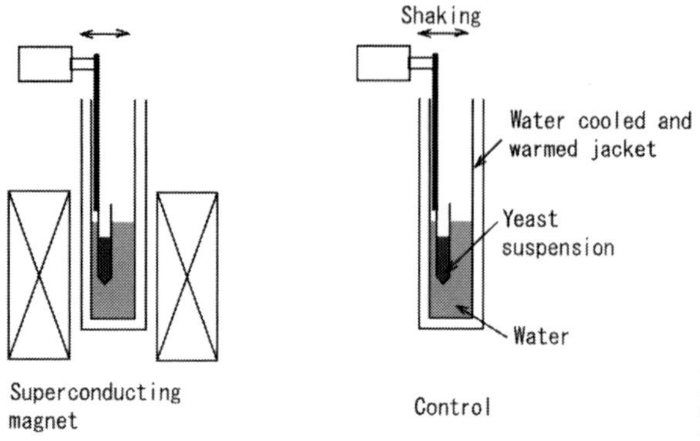

Figure 1. Configuration of sample holder.

2.2. Experimental yeast

The second experiment was performed for experimental yeast. The experimental time was one week in previous sake mash experiment, since the temperature was low and the condition was the stationary cultivation. Therefore, in the second experiment the temperature was controlled to be higher and shake cultivation was used. The yeast used in the second experiment was so-called experimental yeast W303-1a. And the yeast cells were pre-cultured on a YPD (Yeast extract Peptone Dextrose) agar plate. Then it was moved to YPD suspension and 12 hours incubation was performed in a water bath with shaking. Then, the pre-cultured yeast cells were put to 10 ml YPD with the number density of yeast cells of $5.0 \pm 0.2 \times 10^6$/ml. This yeast suspension is divided into two tubes. Here, YPD was composed by Yeast extract: Peptone: Dextrose=5 g: 10 g: 10 g to 500 ml RO water.

The magnetic field was applied by the superconducting magnet up to 10 T. The temperature was controlled by the jacketed vessels in the range of 20–30°C. The tubes of the yeast suspensions were put inside the jacketed vessels as shown in figure 1. The tubes were shaken about 1 Hz to maintain a good condition for the proliferation. The shake stroke was 10 mm resulting in small magnetic-induced force, because the magnetic field is uniform within 1.4% in a sphere with radius of 40 mm. In the shaken condition, since the tubes were open to expose the yeast suspension to the air, oxygen was enough supplied to the yeast cells, and the activity of the yeast cells was accelerated. The position of the yeast suspension was the center of the superconducting magnet resulting in very small effect of the magnetic-induced force by the diamagnetism of oxygen or water and magnetic gradient. And the magnetic field was applied in the range of 0~ 10 T.

The time dependence of the number density of the yeast cells ρ in the suspension was measured. The number density of the yeast cells for magnetic field ρ_M and for control ρ_C were counted by standard plate count technique.

3. Results and Discussion

The result of quantitative analysis for chemical components after about one week alcohol fermentation of Japanese sake mash in the magnetic field of 10 T is listed in Table 1. The differences of samples of magnetic field and control in (a) low-boiling flavor component and (b) organic acid are within 1.6ppm except pyruvic acid. Hence, the effects of magnetic field to flavor component and organic acid are very small. However, the concentration of the glucose in the magnetic field is larger than that in the control, and the concentration of the ethanol in the magnetic field is smaller than that in the control. These results are also confirmed by the organoleptic examination by the members of Fukuoka Regional Taxation Bureau, i.e., three appraisers concluded sweet taste for the sample in the magnetic field. Figure 2 shows the time dependence of the number density of the yeast cells at 10 T and 15°C. It is found that the number density of the yeast cells in the magnetic field, ρ_M, is suppressed than the case of the

Table 1. Results of quantitative analysis for (a) low-boiling flavor component, (b) organic acid, and (c) glucose and ethanol in the magnetic field and control.

	(a) low-boiling flavor component			
	isobutyl alcohol	isoamyl alcohol	isoamyl acetate	ethyl caproate
magnetic field(ppm)	10.4	32.7	0.7	0.9
control(ppm)	9.7	34.2	0.7	1.0

	(b) organic acid					
	succinic acid	malic acid	lactic acid	pyruvic acid	citric acid	acetic acid
magnetic field(ppm)	73.5	86.8	30.7	668.3	104.0	85.9
control(ppm)	71.9	87.8	30.3	648.4	104.8	87.5

(c)	glucose	ethanol
magnetic field(%)	8.1	3.5
control(%)	7.9	3.7

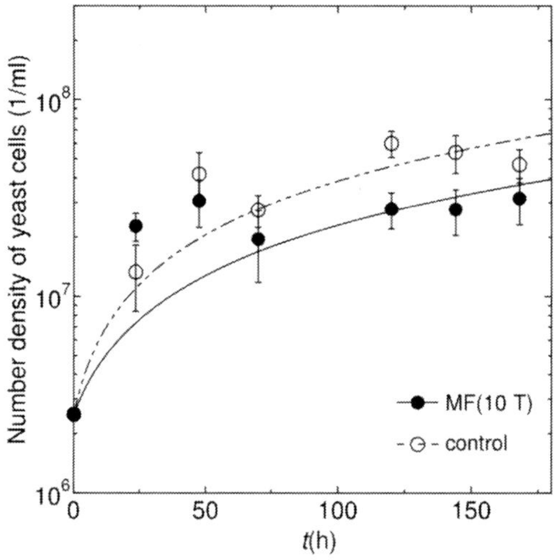

Figure 2. Time dependence of number density of yeast K-7 cells at 10 T and 15°C.

control, ρ_C. These results suggests that the activity of the yeast cells is suppressed by the strong static magnetic field resulting in the small number density of the yeast cells. That is, the glucose is not sufficiently converted to the ethanol in the magnetic field due to the low activity of yeast cells, resulting in sweet taste.

The control of the activity of the yeast cells is mainly done by the control of the temperature in brewing. For example, high quality sake is made in low temperature such as 10–15°C and the fermentation rate is controlled to be lower than usual sake brewing, and particular flavor is produced. If the fermentation rate is controlled by the magnetic field, new kinds of taste could be created in new circumstance with the magnetic field. For example, much slower fermentation could be achieved at low temperature and strong magnetic field. Therefore, it should be confirmed that the effect of the magnetic field to the yeast cells in various magnetic fields and temperatures. In this study, the experimental yeast W303-1a is used for the measurement of the time dependence of the number density of the yeast cells in shirking condition in various conditions.

Figure 3 shows the result of the time dependence of the number density of the experimental yeast W303-1a cells at 10 T and 20°C for shaking condition as shown in figure 1. The number density of the yeast cells is almost saturated at 30 hours, since the nutrition in YPD is limited.

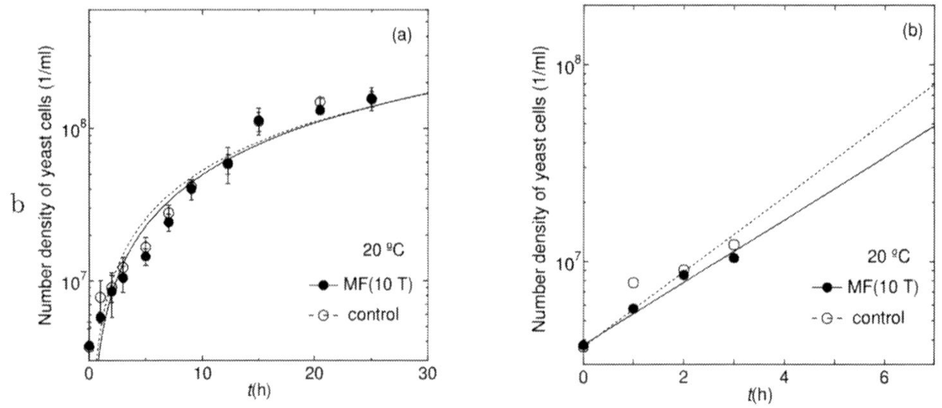

Figure 3. Time dependence of number density of yeast cells at 10 T and 20°C for yeast W303-1a.

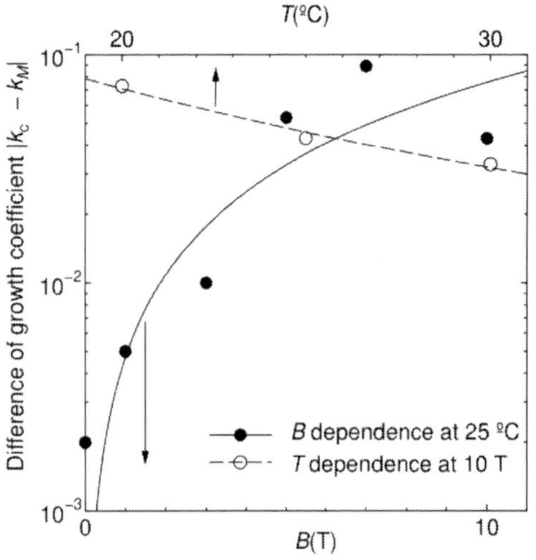

Figure 4. Magnetic field dependence of growth coefficient (solid line) at 25°C and temperature dependence of growth coefficient (dashed line) at 10 T.

The rate of the fermentation of 30 hours is faster than the previous experiment of Japanese sake mash of 7 days, since the temperature is high and there exists much oxygen provided by the shaking condition. It is found that the number density of the yeast cells in the magnetic field is suppressed in the time less than 10 hours. This period is known as most active time in the proliferation of the yeast cells. These results are confirmed in the temperature range of 20–30°C and the magnetic field range of 1–10 T. Reproducibility tests are performed at 25°C and 5 T and almost the same results are obtained.

The rate of the proliferation is tried to be defined in the most active time less than 6 hours. The time dependence of the number density of the yeast cells, ρ, is analyzed in terms of Malthus equation as given by

$$\rho = \rho_0 \exp(kt). \tag{1}$$

In the above, k is the growth coefficient and ρ_0 is a constant indicating the initial number density of the yeast cells at initial time $t = 0$. The example of determination of k is shown in figure 3(b). It is found that the growth coefficients k of two cases shown as the slopes in the figure are different and the k for the case of the magnetic field k_{M} is smaller than that of the control k_{C}. The difference of the growth coefficient $|k_{\mathrm{C}} - k_{\mathrm{M}}|$ of the magnetic field, k_{M}, and the control, k_C is shown as functions of magnetic field and temperature in figure 4. The temperature is 25°C for the magnetic field dependence, and the magnetic field is 10 T for the temperature dependence. The difference of the growth coefficients becomes small as increasing temperature, since the activity of the yeast cells is high at higher temperatures and the effect of the magnetic field is less remarkable. It is also found that the effect of magnetic field for the growth coefficient increases as increasing the magnetic field. However, it seems to be saturated around 5 T. Therefore, it is necessary to confirm the effect of the magnetic field on the yeast

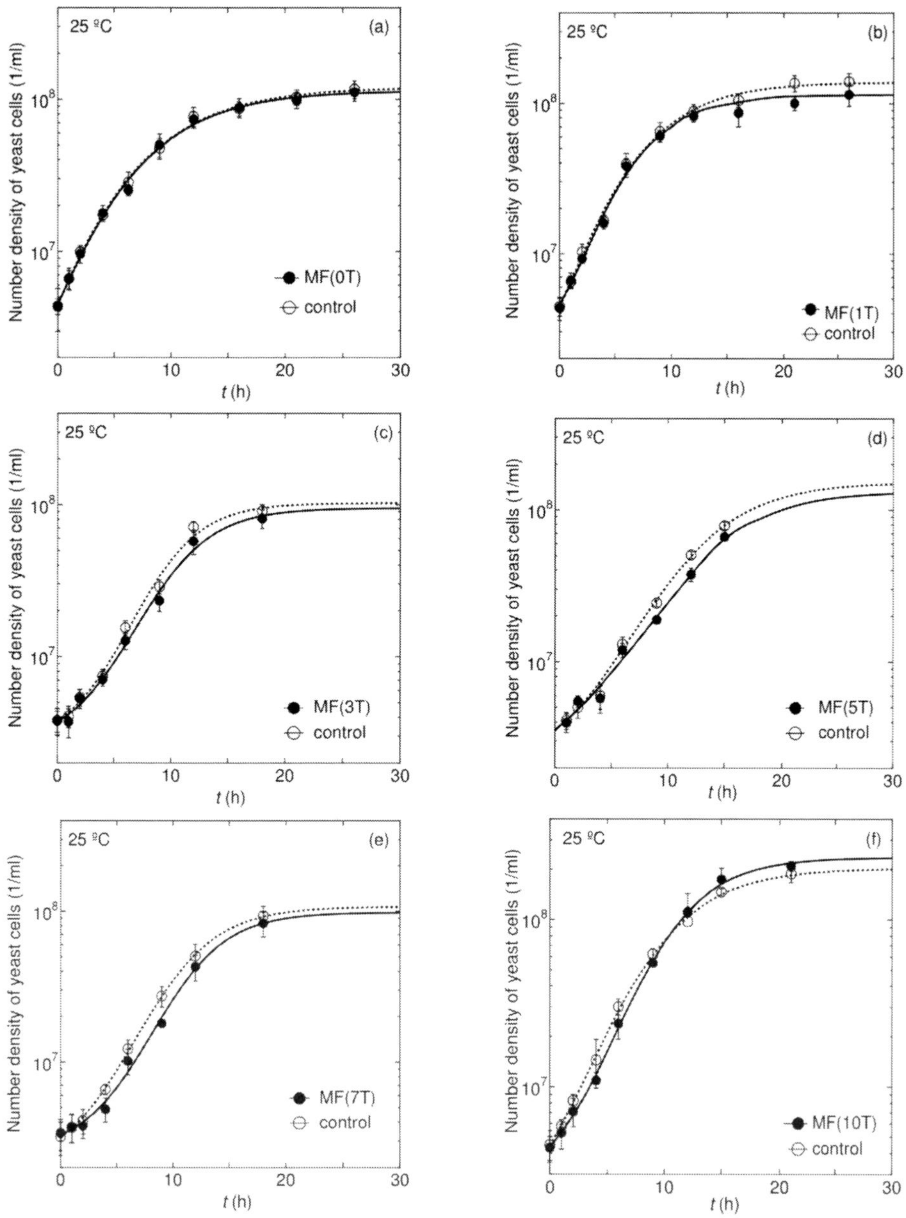

Figure 5. Time dependence of number density of yeast cells for various magnetic fields at 25°C, (a) 0 T, (b) 1 T, (c) 3 T, (d) 5 T, (e) 7 T, (f) 10 T.

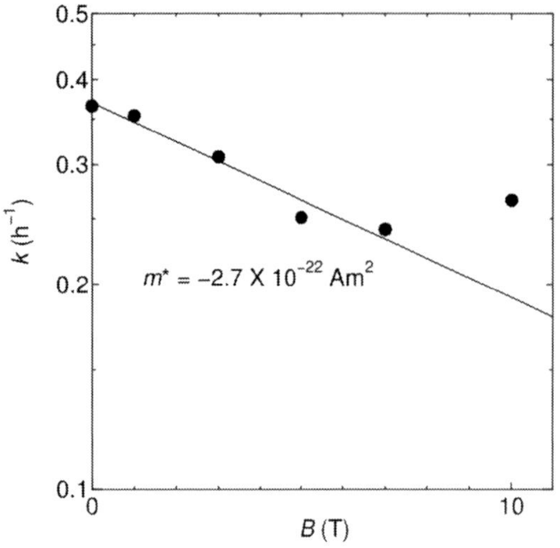

Figure 6. Magnetic field dependence of the growth coefficient, k at 25°C.

cells activities at stronger magnetic fields higher than 10 T. The time dependence of the number density of the yeast cells at 25°C with various magnetic field are shown in figure 5. The difference is enough small at 0 T, indicating the lower limit of detecting the difference of k in the present experiment. The difference of the growth coefficient becomes large as increasing magnetic field and saturates around 5 T as abovementioned in figure 4.

Here, the growth coefficient is discussed by using classical Arrhenius type equation which is frequently used in physicochemical investigation. The magnetic field dependence of the growth coefficient is described by

$$k(B) = k_0 \exp\left(\frac{m^*}{k_B T} B\right), \tag{2}$$

where k_B is the Boltzmann constant and m^* is the magnetic growth activation moment which represent the sensitivity to magnetic field[8]. That is, large value of m* means that the growth activation of yeast cells is sensitive by the magnetic field. Figure 6 shows the semi-logarithmically plot of magnetic field dependence of the growth coefficient. Hence, the magnetic growth activation moment is determined from the initial slope as $m^* = -2.7 \times 10^{-22}$ Am^2. It is desired to measure m^* for various cases such as various temperatures or various kinds of yeast cells.

It is difficult to know the mechanism for the strong static magnetic field effect of the yeast cells from the present experiment. However it is interesting to know that the effect of the magnetic field is not related to the joule heat since the magnetic field is static, and is not related to the magnetic-induced force by water or oxygen since the magnetic gradient is negligibly small in the present study. It should be noted that the effect of the oxygen is small in the experiment with the experimental yeast, since it is shaking condition and this physical effect is much larger than the magnetic-induced force to the oxygen. It is reported that the vaporization of water is enhanced by gradient magnetic field more than 10 T^2/m[9]. However, this effect seems to be small since the tubes is vertically placed in the center of the superconducting magnet and the gradient magnetic field is small. Therefore, it is desired to investigate the essential explanation of the magnetic field on the activities of the yeast cells.

4. Summary

In this study, the effect of strong static magnetic field to the activities of the yeast cells were measured in various magnetic fields and temperatures. As a result, the proliferation of the yeast cells is suppressed by applying the static magnetic field. Therefore, the taste of sake mash becomes sweet when magnetic field is applied during the fermentation process. The effect of the magnetic field is saturated around 5 T. When the temperature is high, the effect of the magnetic field becomes small, since the proliferation activity is high at high temperatures. It is considered that the control of the fermentation of the yeast cells is possible by the strength of the magnetic field in addition of the control of the temperature. Therefore, the new taste of alcoholic beverage is expected with applying static magnetic field. For example, much slower fermentation could be achieved at low temperature with strong magnetic field.

Acknowledgments
We thank Mr. and Mrs. K. Yano of Tamanoi Shuzo Co. Ltd. for helpful discussions. We thank Dr. E. Neumann of Bielefeld University for his editorial advice in data presentation and physical chemical analysis.

References

[1] Novák J, Strašák L, Fojt L, Slaninová I and Vetterl V 2007 Effects of low-frequency magnetic fields on the viability of yeast Saccharomyces cerevisiae *Bioelectrochemistry* **70** 115–121
[2] Ruiz-Gómez M J, Prieto-Barcia M I, Ristori-Bogajo E and Martínez-Morillo M 2004 Static and 50 Hz magnetic fields of 0.35 and 2.45 mT have no effect on the growth of Saccharomyces cerevisiae *Bioelectrochemistry* **64** 151–155
[3] Watanabe K, Awaji S, Motokawa M, Mikami Y, Sakuraba J and Watazawa K 1998 15 T Cryocooled Nb_3Sn Superconducting Magnet with a 52 mm Room Temperature Bore *Jpn. J. Appl. Phys.* **37** L1148–L1150
[4] Horiuchi S, Ishizaki Y, Okuno K, Ano T and Shoda M 2001 Drastic high magnetic field effect on suppression of Escherichia coli death *Bioelectrochemistry* **53** 149–153
[5] Zhang Q -M, Tokiwa M, Doi T, Nakahara T, Chang P -W, Nakamura N, Hori M, Miyakoshi J and Yonei S 2003 Strong static magnetic field and the induction of mutations thought elevated production of reactive oxygen species in Escherichia coli soxR *Int. J. Radiat. Biol.* **79** 281–286
[6] Ikehata M, Iwasaka M, Miyakoshi J, Ueno S and Koana T 2003 Effects of intense magnetic fields on sedimentation pattern and gene expression profile in budding yeast *J. Appl. Phys.* **93** 6724–6726
[7] Iwasaka M, Ikehata M, Miyakoshi J and Ueno S 2004 Strong static magnetic field effects on yeast proliferation and distribution *Bioelectrochemistry* **65** 59–68
[8] Neumann E 2000 Digression on chemical electromagnetic field effects in membrane signal transduction — cooperatively paradigm of the acetylcholine receptor *Bioelectrochemistry* **52** 43–49
[9] Nakagawa J, Hirota N, Kitazawa K and Shoda M 1999 Magnetic field enhancement of water vaporization *J. Appl. Phys.* **86** 2923–2925

Journal of Physics: Conference Series **156** (2009) 012017 doi:10.1088/1742-6596/156/1/012017

Effects of Gradient Magnetic Force and Diamagnetic Torque on Formation of Osteoclast-like Giant Cell

M Iwasaka[1], M Ikehata[2] and N Hirota[3]

[1] Department of Medical System Engineering, Graduate School of Engineering, Chiba University, Chiba 263-8522, Japan

[2] Railway Technology Institute, 2-8-38 Hikari-cho, Kokubunji-shi, Tokyo 185-8540, Japan

[3] National Institute for Materials Science, Sengen1-2-1, Tsukuba 305-0047, Japan

E-mail: iwasaka@faculty.chiba-u.jp & iwasaka-m@umin.ac.jp

Abstract. In bone tissue, two kinds of cells, osteoblast (OB) and osteoclast (OC), contribute to remodeling of bone. In the present study, a co-culture system of bone-forming cell (OB) and -dissolving cell (OC) was incubated in static magnetic fields of horizontal 14 T and vertical gradient 10 T. Effect of two kinds of magnetic fields was an inhibition of OC formation. Three kinds of mechanisms, magnetic orientation of OB, diamagnetic torque force acting on OC, and possible reduction of earth's gravity were discussed.

1. Introduction

Bone tissue, which is in a solid state in the living body, also contains both positive and negative control processes. Two kinds of cells, osteoblast (OB) and osteoclast (OC), contribute to the formation and deformation of bone, respectively. In the present study, a co-culture system of bone-forming and -dissolving cells was incubated in a static magnetic field at 14 Tesla (T), and effects of magnetic orientation of OBs on the increase of the OCs that erode bones were investigated.

Previous studies indicated that dia-magnetism, which is an induced magnetism of all materials under magnetic fields, possibly affect biological cells [1-4]. It is expected that two kinds of magnetic forces, diamagnetic torque force and gradient magnetic force possibly affect a cell component in living cell [1-3,5,6].

The diamagnetic torque force has effects on the rotational motion of molecules, and potentially affects the morphology of cells. The utilized superconducting magnet has gradient magnetic forces in a direction of bore's axis in the superconducting coil. The magnetic flux provides torque forces on the fibers that assemble to form the cells [1, 5, 6].

We utilized an experimental system to investigate the effects of magnetic fields on bone formation. The experimental system is the co-culture system with both OB and OC. The processes start with bone marrow cells, and the cells differentiate to precursor OC, OB, etc.. The OC precursors do cell fusions with an assistance of OB and other factors, such as receptor activator of NF-κB ligand (RANKL) and interleukin 1 (IL-1) [7-9]. Finally, they manifest a large sized multi-nucleated cell, OC-like cell.

Our purpose is to apply these two kinds of magnetic force to control of living cells. The diamagnetic control of living cells was investigated by using bone cells. Bone has a solid state, which is known as hydroxy-apatite, in living body. The hydroxy-apatite is degraded by OCs, and reconstructed by OB to exhibit a bone remodeling [10].

Our previous studies showed that the magnetic orientation of OBs occurred under the horizontal magnetic fields of up to 8 T [11], however, we have still not obtained any data about magnetic field effect on OCs. As a model for the magnetic field effects on bone remodeling in vitro, a co-culture system with both OB and OC was exposed to a strong static magnetic field of up to 14 T.

2. Methods

2.1. Cell culture system for co-culture with OB and OC

The culture of bone marrow cells was carried out by using a cell culture dish (100mm in diameter, CORNING, USA) and a slide flask (17920, NUNC Co., Ltd., USA).

The right photo of Fig. 1 shows an OC, which was observed after 5 days of bone marrow incubation (in the absence of magnetic field exposure). The OC was surrounded by fibroblastic cells, OBs. The observed OC-like cell had about ten nuclei.

2.2. Magnetic field generators

We utilized three kinds of superconducting magnets, with a horizontal bore (max. 14 T) and with a vertical bore (max. 12T). The temperature in magnet bore was thermally stabilized by a cylindrical type of water jacket. In case with the cell culture in magnet during several days, the incubation was carried out without CO_2 control.

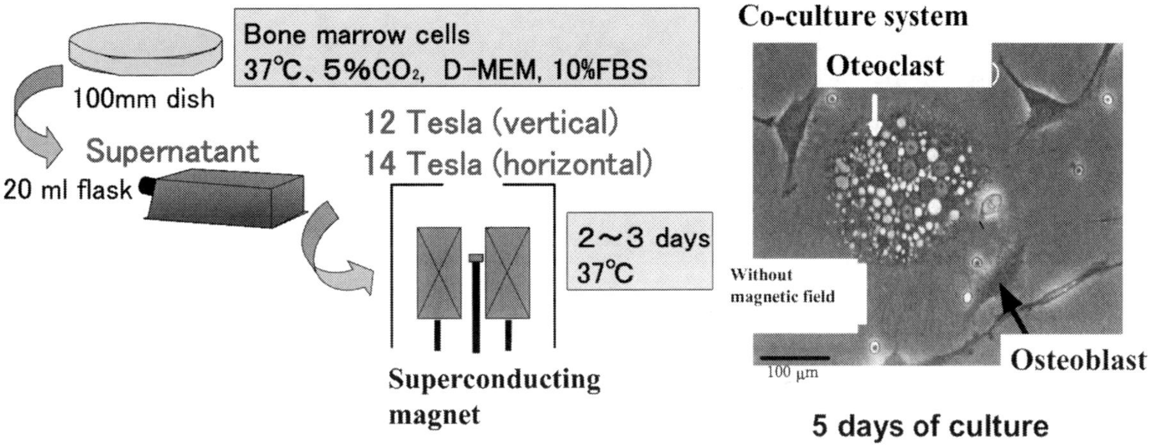

Figure 1. Experimental procedure for the differentiation of bone marrow cells to OC-like giant cell and OB. Right photo shows an example of multi-nucleated giant cell (OC), which is surrounded by OB.

Journal of Physics: Conference Series **156** (2009) 012017 doi:10.1088/1742-6596/156/1/012017

3. Results and Discussion

3.1. Horizontal magnetic field effect

After three to five days of primal cell culture, the cells were moved to a flask for magnetic field exposure. Then the cells were cultured with and without magnetic fields of 14 T for 50 hours (about two days).

An effect of magnetic fields, 14 T, was observed in the bottom photo of Fig. 2. OBs oriented parallel to the applied magnetic field direction. Top photo shows an OC with dozens of nuclei. The OB orientation near the OC was random while the orientation of OBs apart from OC oriented parallel to the magnetic fields.

Concerning the effects of magnetic fields on morphologies of OC, we compared the size and density of OCs under magnetic fields with those of OCs under ambient fields. An example of effect of magnetic fields on the size of OC is shown in Fig. 3. After starting the co-culture of OB and OC, matrix under OB caused contractions. Then the area of OC was separated from OBs, and OCs were observed in OB-poor area.

A normal circle pattern of OC formed a rounded area where OBs were absent when the sample was incubated for two days without the magnetic field exposure (left panel of Fig. 3). In case with magnetic field exposure at 14 T for two days, it was frequently observed that the pond became ellipsoidal and the size of OC decreased, as shown in an example in Fig. 3 (right panel).

Figure 4 shows the size parameters, the length of long axis (L) and short axis (S), of OC cells with and without magnetic fields of 13~14 T. The results showed that the magnetic fields decreased the size of multi-nuclei-area of OC. It was suggested that the 14 T horizontal magnetic fields decelerated the formation of OC-like multi-nucleated cells when the OBs were exposed to the magnetic fields after the cell passage.

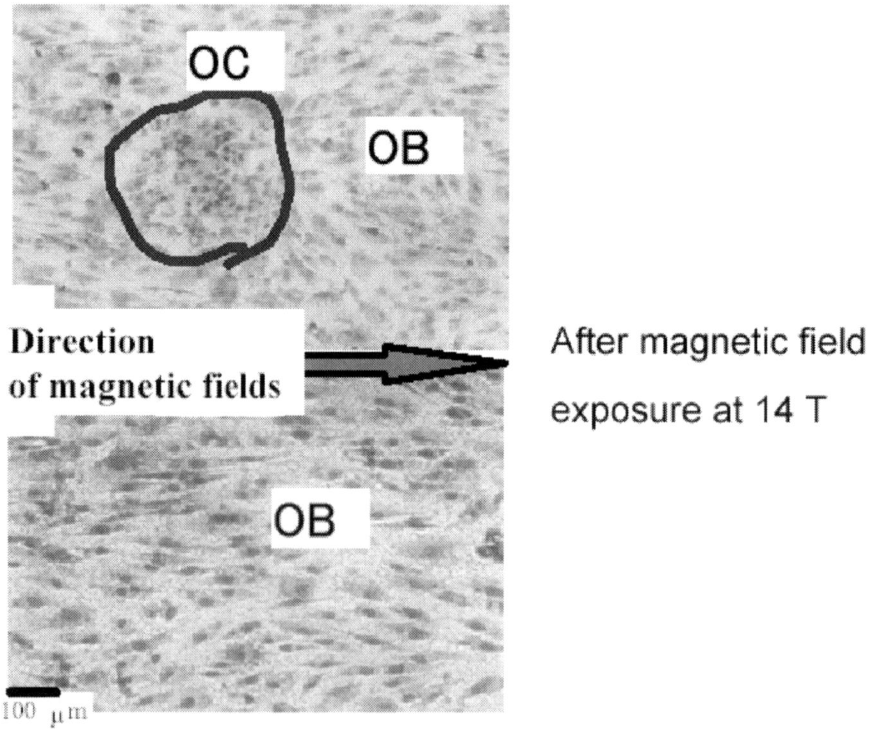

Figure 2. Effects of horizontal magnetic fields on OB and OC. The cells were stained with Gimsa's solution after fixation with methanol.

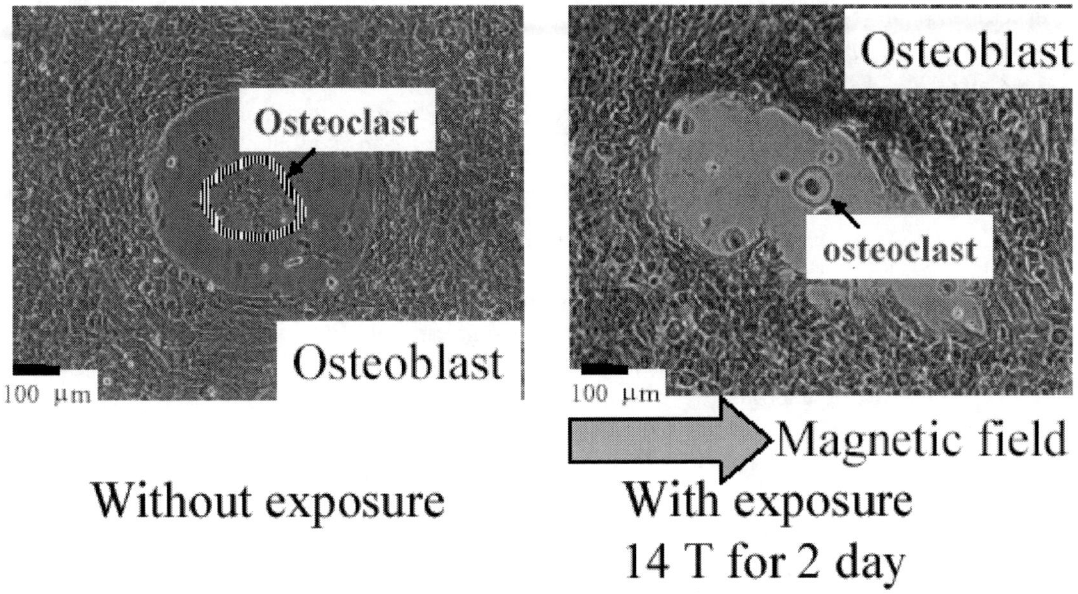

Figure 3 . Effects of horizontal magnetic fields on OC and OB.

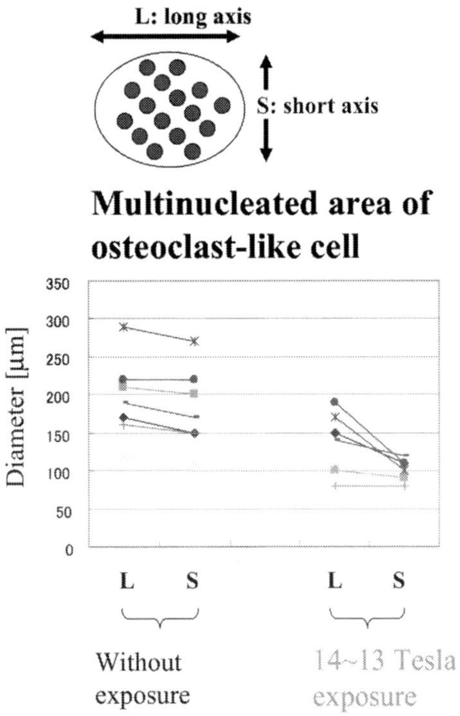

Figure 4. Effects of magnetic fields of 13~14 T on size of OC-like multi-nucleated giant cells.

3.2. Vertical magnetic field effect

In the next, we carried out the OC-OB co-culture under vertical gradient magnetic fields of 10 T which was generated at an upper region of the superconducting magnet's bore. The co-culture procedure was same as the experiment with horizontal fields except exposure period. In the early stage of culture, the cells were exposed to the 10 T vertical fields for three days, and continuously incubated under environmental fields (~0 T) of inside CO_2 incubator for more three days.

The observed effects of a vertical magnetic flux and a vertical gradient force in parallel to the direction of gravity, was an inhibition of multi-nucleation rate in the process of OC formation, as shown in Figure 5. The two top panels (photos) show that the 10 T magnetic field exposure exhibited an OC with a size smaller than the control. We measured both long and short axes of OC and obtained a histogram of mean radii OC (bottom panels). The 10 T gradient fields increased the number of small OCs with maximum mean radii of less than 100 microns while reduced the number of OCs in the mean size of more than 200 microns.

The data indicated a possibility that the magnetic field exposure condition inhibited a cell fusion of pre-cursor OC and consequently had a delay in forming a multi-nucleated giant cell.

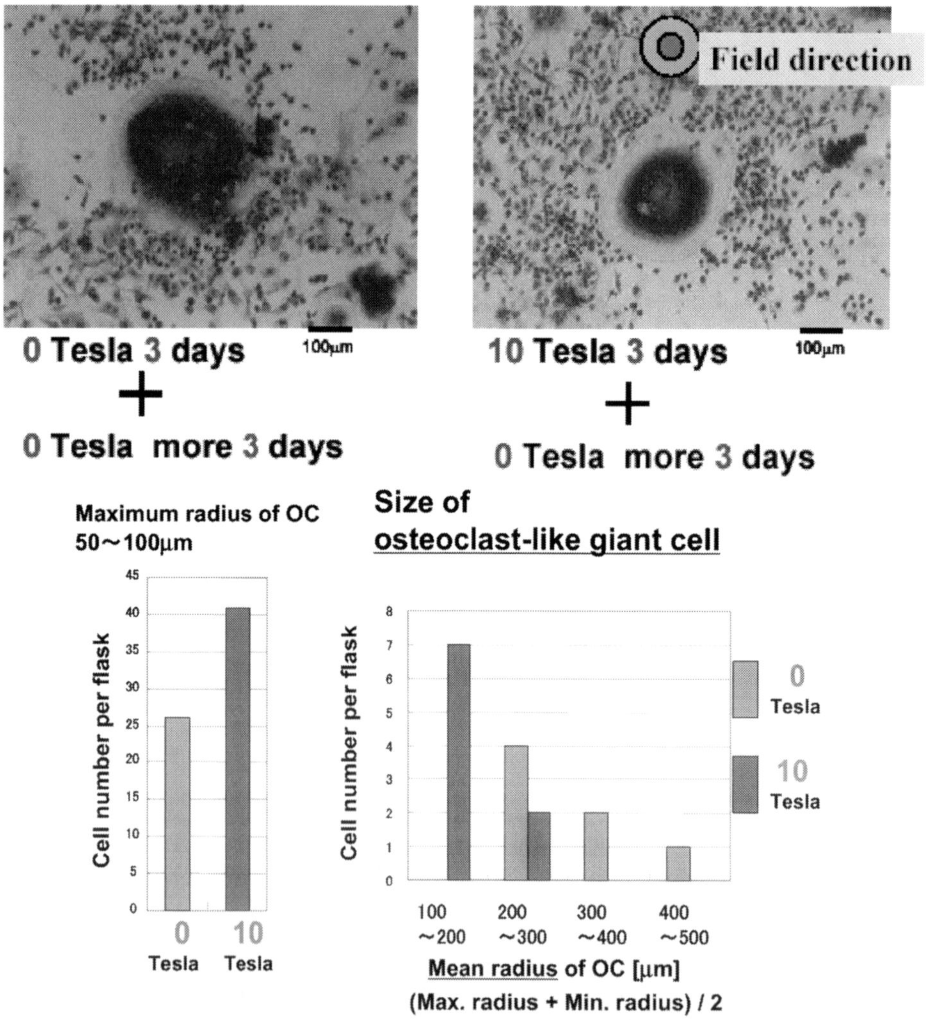

Figure 5. Effect of vertical magnetic fields on OC formation.

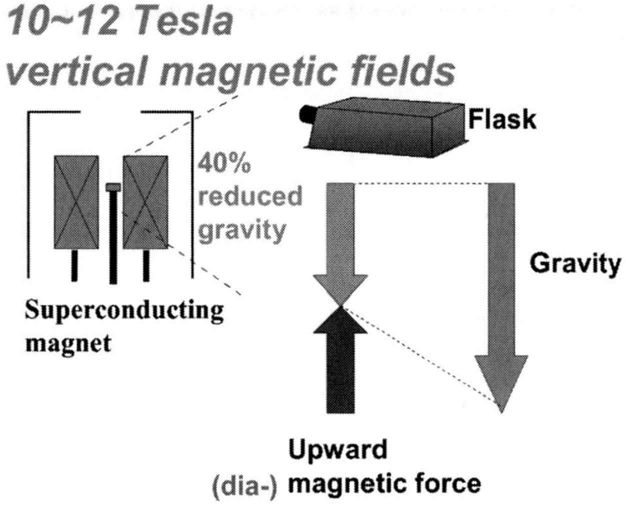

Figure 6. An illustration of magnetically reduced gravity which was equivalently produced for diamagnetic molecules.

3.3. Possible mechanisms and application

The results suggested that the suppression of OC size occurred both under horizontal and vertical magnetic fields of 10T order. An optional parameter, reduced gravity by diamagnetic force, was added to the experiments in the vertical bore (12 T vertical magnet).

In general for the formation of a multi-nucleated giant cell via a cell fusion process, cell to cell interaction has an important effect on the cell fusion. Particularly in the formation of OC, precursor OC to OB interaction was also considered to assist OC growth [9]. We speculated that a cell to cell interaction between growing OC and magnetically orienting OB had an important role in forming OCs under horizontal magnetic fields. We observed that the OBs in an appropriate cell density caused magnetic orientation, and its consequence, cellular assembly pattern changes where no OC exists. However, the presence of OC provided an agitation or confliction to the OB cell assembly, and probably the diamagnetic torque forces acting on OBs were accumulated in a cell assembly, then the diamagnetic anisotropic energy was applied for a suppression of cell fusion. No extension of OC morphology was observed under the 14 T horizontal magnetic fields.

In case with the OC-OB co-culture under vertical gradient magnetic fields of 10 T, the earth's gravity was reduced by about 40 %. As shown in Figure 6, the utilized superconducting magnet with a vertical bore generated a maximum magnetic field of 12 T at a center of bore, and the upper side of gradient fields provided a diamagnetic force which was anti-parallel to earth's gravity force, and as a result, a diamagnetic object was exposed to a field equal to a reduced gravitational field. The mechanism was not expected in the experiments of horizontal magnet. Possibly the reduced gravity by gradient magnetic fields was one of the mechanism candidates. To clarify its contribution to the OC formation quantitatively, further study under an enhanced gravity mode, at the lower part of magnet's bore should be carried out in the future.

In the cells being exposed to gradient magnetic fields, a diamagnetic torque force was superposed on gradient magnetic forces. Previous our study suggested that the vertical magnetic fields changed the alignment of lipid molecules in cell membrane and the cellular morphology became balloon-like. The mechanism also resulted in making the size of OC small.

Multiplicate mechanisms existed in the observed phenomenon and above mentioned three mechanisms can make a delay in OC formation; i.e. cellular fusion. Preventing OC formation in a bone tissue of living body causes an enhancement on calcification and hydroxy apatite generation in bone tissue. So it is considered that 10 T order magnetic fields are useful for controlling bone metabolism.

4. Conclusions

Horizontal magnetic fields of 14 T induced an inhibition OC formation by means of cellular fusion. A magnetic orientation of OBs was observed in an OB rich area.

Vertical gradient magnetic fields at 10 T also induced an inhibition OC formation. Cell differentiation and cellular fusion process had a delay due to the magnetic field exposure.

Acknowledgement

This work was partly supported by Kanzawa Medical Research Foundation.

References

[1] Valles J M Jr, Wasserman S R, Schweidenback C, Edwardson J, Denegre J M, Mowry K L 2002 *Exp Cell Res.* **10** 274 112

[2] Higashi T, Yamagishi A, Takeuchi T, Kawaguchi N, Sagawa S, Onishi S, Date M 1993 *Blood* **82** 1328

[3] Umeno A, Kotani H, Iwasaka M, Ueno S 2001 *IEEE Trans. Magn.* **37** 2909

[4] Fujiwara Y, Tomishige M, Itoh Y, Fujiwara M, Shibata N, Kosaka T, Hosoya H, Tanimoto Y 2006 *Molecular Physics* **104** 1659

[5] Iwasaka M, Ueno S, 2003 *Bioelectromagnetics* **24** 564

[6] Iwasaka M, Miyakoshi J, Ueno S 2003 *In Vitro Cellular & Developmental Biology – Animal* **39** 120

[7] Linkhart T A, Linkhart S G, Kodama Y, Farley J R, Dimai H P, Wright K R, Wergedal J E, Sheng M, Beamer W G, Donahue L R, Rosen C J, Baylink D J 1999 *J. Bone Miner. Res.* **14** 39

[8] Kondo Y, Irie K, Ikegame M, Ejiri S, Hanada K, Ozawa H 2001 *J. Bone Miner. Metabol.* **19** 352

[9] Mutsuzaki H, Sakane M, Ito A, Nakajima H, Hattori S, Miyanaga Y, Tanaka J, Ochiai N 2005 *Biomaterials* **26** 1027

[10] Blair H C 1998 *Bioessays* **20** 837

[11] Kotani H, Kawaguchi H, Shimoaka T, Iwasaka M, Ueno S, Ozawa H, Nakamura K, Hoshi K 2002 *J. Bone Miner. Res.* **17** 1814

Journal of Physics: Conference Series **156** (2009) 012018 doi:10.1088/1742-6596/156/1/012018

X-ray Powder Diffraction Camera for High-Field Experiments

K Koyama, Y Mitsui, K Takahashi and K Watanabe

High Field Laboratory for Superconducting Materials, Institute for Materials Research, Tohoku University, Sendai 980-8577, Japan

E-mail: kkoyama@imr.tohoku.ac.jp

Abstract. We have designed a high-field X-ray diffraction (HF-XRD) camera which will be inserted into an experimental room temperature bore (100 mm) of a conventional solenoid-type cryocooled superconducting magnet (10T-CSM). Using the prototype camera that is same size of the HF-XRD camera, a XRD pattern of Si is taken at room temperature in a zero magnetic field. From the obtained results, the expected ability of the designed HF-XRD camera is presented.

1. Introduction

Recently, high performance materials controlled by magnetic fields such as ferromagnetic shape memory alloys [1,2], giant magnetostrictive compounds [3], magnetic refrigerants [4-7] and so on have been intensively studied all over the world. In these compounds, a magnetic field plays an important role for application-oriented research. Many compounds of the above system exhibit a structural transformation or a large lattice distortion, accompanied by the magnetic phase transition at the working temperature. Therefore, it is very important for developing these materials to clarify structural properties microscopically in magnetic fields. Especially, it requires high magnetic fields to investigate the potential of these materials at the early stage of the development.

So far, neutron and synchrotron X-ray diffraction experiments have been performed in high magnetic fields over 15 T using split-pair type superconducting magnets [8] and 25 T using pulsed magnets [9, 10], respectively. In addition, a 30 T hybrid magnet has been also planned for neutron experiments at the Hahn-Meitner Institute, Germany, collaborating with the National High Magnetic Field Laboratory, USA [11]. However, it is difficult to investigate many materials in detail using such big facilities, because of limited machine time and experimental cost.

An X-ray powder diffraction (XRD) experiment in house is a very general and useful method to investigate the structural properties of materials. Some researchers have independently reported compact X-ray diffraction systems using split-pair type superconducting magnets [12-14]. In the High Field Laboratory for Superconducting Materials, Institute for Materials Research, Tohoku University (HFLSM), the high-field X-ray powder diffraction (5T-HF-XRD) apparatus has been developed [13]. This apparatus consists of a cryocooled split-pair superconducting magnet and θ -2 θ scan-type X-ray diffractometer. The apparatus has provided unique experiments in magnetic fields B up to 5 T and 8 – 320 K temperature T range. By using this system, several important results have been reported up to now [15-17]. However, it is difficult to upgrade an X-ray diffractometer combined with high field magnets over 10 T. Development of a split-pair superconducting magnet, which can generate more than 10 T with a horizontal room temperature gap for X-ray beam path, has a lot of problems like a mechanical strength issue to be solved and needs a huge cost to construct.

In this work, we have started to develop an X-ray diffraction apparatus in high field measurements. A diffractometer is to be inserted into an experimental room temperature bore (100 mm) of a magnet. In this case, we abandoned the conventional diffractometer method but adopted the Debye-Scherrer camera, which is simple and not influenced by the magnetic field. However, it is necessary to examine whether the experimental conditions are realistic in high magnetic fields, before developing the camera system. Furthermore, it should be carefully examined that the camera system with a power supply, an X-ray source, the controlling device, and safety devices will work in large leakage fields (~0.1 T) of the high field magnet. This paper describes a basic design of a HF-XRD camera system and some results obtained using its prototype camera in a zero magnetic field. The obtained results suggest that the XRD pattern is obtained in high magnetic fields.

2. Design of a high-field X-ray powder diffraction camera system

Figure 1 shows the schematic illustration of the high-field X-ray powder diffraction (HF-XRD) camera system at present, which is utilized for a 10T-100 mm bore cryocooled superconducting magnet (10T-CSM). The apparatus mainly consists of a Debye-Scherrer camera (TRY SE Co. Ltd), an X-ray source, a power supply, a chiller and 10T-CSM.

An X-ray source is set outside from the magnet end, which is 472 mm apart from the center (sample position) of the magnet. We adopt a vacuum-tube type X-ray source. Therefore, we can easily change the wavelength by selecting a target among Cu, W, Fe, Co and Mo. The maximum output power of an X-ray source is 2 kW. A collimator size is ø 1.0 or 0.5 mm. The X-ray pass from the source to the collimator is evacuated by a vacuum pump.

A diffraction pattern is collected on an image plate (IP) detector in the Debye-Scherrer camera. The pixel size of the IP leader will be 0.025 mm × 0.025 mm. The diameter of the camera is 80.2 mm. A measurement range in 2θ is from -140° to 140°. A powder sample is set in a capillary (ø 0.5 mm) on the center axis of the camera. In our system, the IP can be changed with the sample fixed in high magnetic fields.

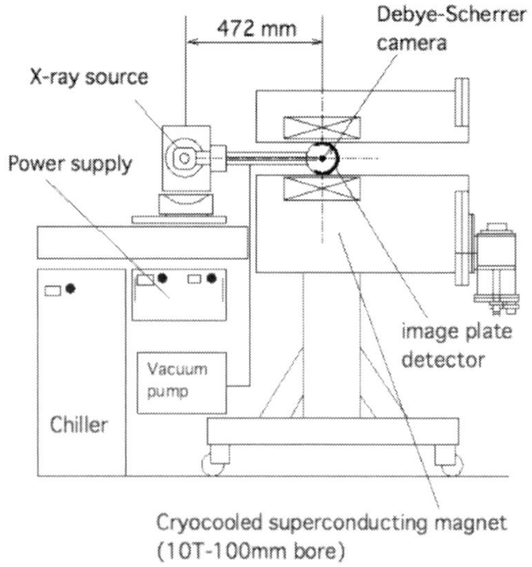

Figure 1. Schematic illustration of the high-field X-ray powder diffraction (HF-XRD) camera system which is combined with a 10T-100 mm bore cryocooled superconducting magnet (10T-CSM).

3. Estimation of the performance of HF-XRD using a prototype camera

Before installing the HF-XRD camera in 10T-CSM, we estimated the performance of the camera using a prototype one without the magnet. The diameter (80.2 mm) of the prototype camera is the same size of the HF-XRD camera. Figure 2 shows the XRD pattern of Si powders at room temperature in a zero magnetic field, which is collected by the IP detector (Fujifilm BAS III) using the prototype camera. The obtained pattern can be indexed, as shown in this figure. This measurement is performed in air using Cu Kα radiation (1.2 kW), a collimator (ø 1.0), a Ni filter and IP. The X-ray beam path from the source to sample is 150 mm, and the exposure time is 15 minutes.

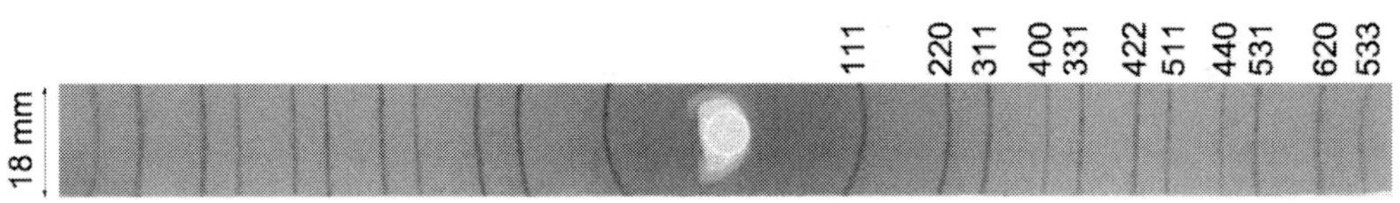

Figure 2. X-ray powder diffraction pattern of Si at room temperature in a zero magnetic field. The pattern is collected by an image plate detector using a prototype camera that is same scale of the HF-XRD camera.

Figure 3 shows the 400 and 331 diffraction patterns, which is a close-up of the pattern in Fig. 2. We confirmed that the pixel size is 0.1 mm × 0.1 mm, indicating that the 2θ–position is determined by the resolution of 0.13°. As we know, the pixel size is limited by an IP-reader. In this measurement, we utilized an IP-reader (TRY-XIA). However, we will use another high resolution IP-reader with the pixel size of 0.025 mm × 0.025 mm in our HF-XRD camera system. Therefore, we will determine the diffraction position with 0.033° on 2θ unit, indicating that lattice parameters can be determined with high significant figures.

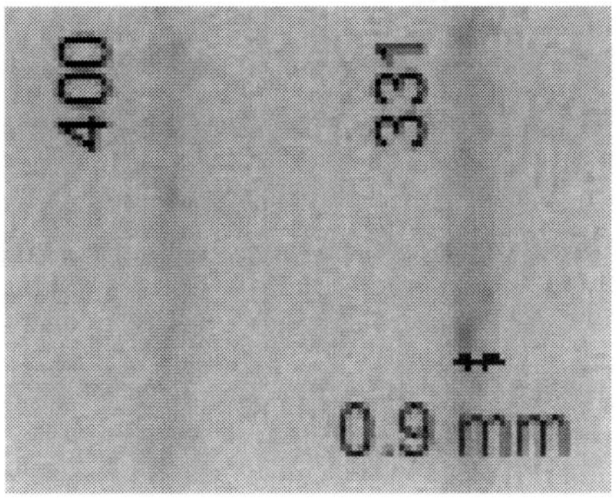

Figure 3. 400 and 331 diffraction patterns, which are close-up of the pattern in Fig. 2. A pixel size is 0.1 mm × 0.1 mm.

The beam pass of the HF-XRD camera is about three times larger than that of the prototype one, indicating that the intensity of the X-ray beam decreases to about 10%. From this result, we roughly estimated that the exposure time is required about 60 minutes for one picture when the output power of the X-ray source is 2 kW and the X-ray pass is evacuated in the HF-XRD camera. This experimental condition is realistic, and the systematic measurement is sufficiently possible in high magnetic fields.

As seen in Fig. 3, the line width of the 331 diffraction is 0.9 mm, which corresponds to 1.2° on 2θ unit. This is two times larger than that obtained by our 5T-HF-XRD. However, it is expected to observe the structural transformation, because the XRD patterns are drastically changed. If we utilize the collimator of ø 0.5 mm, the line width decreases further. From these results, we expect that field-induced structural transformation can be observed using the HF-XRD camera.

We should consider influence of leakage fields from the magnet on the HF-XRD camera. We measured the leakage field distribution of 10T-CSM. The leakage field at 472 mm away from the magnet center is less than 0.07 T for the center field of 10 T. In the case of our 5T-HF-XRD apparatus, an X-ray tube is in the leakage field of 0.08-0.02 T and works well without a magnetic shield in such a condition. Therefore, it is expected that no trouble will be caused on X-ray source by the leakage field in this design, but we also consider the preparation of the magnetic shield for the X-ray source of HF-XRD camera.

From these examinations using the prototype camera, the HF-XRD camera was produced. Figure 4 shows a photo of the HF-XRD camera system that is combined with the 10T-CSM. The HF-XRD camera system examination in high magnetic fields is now in progress. The results will be published in elsewhere in the near future.

Figure 4. Overall view of the high-field X-ray powder diffraction (HF-XRD) camera system that is combined with a 10T-100 mm bore cryocooled superconducting magnet (10T-CSM).

4. Summary

A High-field X-ray powder diffraction (HF-XRD) camera was designed in order to carry out HF-XRD experiments in 10 T generated by a cryocooled superconducting magnet (10T-CSM) at a laboratory site. Using the prototype camera that is the same scale of the HF-XRD camera, we measured a XRD pattern of Si at room temperature in a zero magnetic field. From the results, we expected that one picture of a XRD pattern will be taken about 60 minutes using the HF-XRD camera in 10 T. In addition, obtained results indicate that the high resolution HF-XRD pattern is obtained and the structural transformation can be observed.

Acknowledgments

This work was supported by the Nano-Materials Functionality Creation Research Project for Young Scientists of Institute for Materials Research, Tohoku University.

References

[1] Ullakko K, Huang J K, Kantner C and O'Handley R C 1996 *Appl. Phys. Lett.* **69** 1966
[2] Kainuma R, Imano Y, Ito W, Sutou Y, Morito H, Okamoto S, Kitakami O, Oikawa K, Fujita A, Kanomata T and Ishida K 2006 *Nature* **439** 957
[3] Fujita A and Fukamichi K 1999 *IEEE Trans. Magn.* **35** 3796
[4] Pecharsky V K and Gschneidner Jr. K A 1997 *Phys. Rev. Lett.* **78** 4494
[5] Tegus O, Brück E, Buschow K H J and de Boer F R, 2002 *Nature* **415** 150
[6] Wada H and Tanabe Y 2001 *Appl. Phys. Lett.* **79** 3302
[7] Fujita A, Fujieda S, Fukamichi K, Yamazaki Y and Iijima Y 2002 *Mater. Trans.* **43** 1202
[8] Prokes K, Meissner M, Smeibidl P, Fritsche C, Ohloff K D and Daniels P 2001 *Physica* B **294-295** 691
[9] Nojiri N, Takahashi K, Fukuda T, Fujita M, Arai M and Motokawa M 1998 *Physica* B **241-243** 210
[10] Vanacken J, Frings P, Detlefs C, Duc F, Lorenzo J E, Nardone M, Billette J, Zitouni A, Bras W, and Rikken G, 2006 *J. Phys.: Conf. Ser.* **51** 475
[11] Steiner M, Tennant D A and Smeibidl P 2006 *J. Phys.: Conf. Ser.* **51** 470
[12] Ohsumi H, Tajima K, Wakabayashi N, Shinoda Y, Kamishima K and Goto T 1997 *J. Phys. Soc. Jpn.* **66** 1896
[13] Watanabe K, Watanabe Y, Awaji S, Fujiwara M, Kobayashi N and Hasebe T 1998 *Adv. Cryog. Eng.* **44** 747
[14] Holm A P, Pecharsky V K, Gschneidner Jr K A, Rink R and Jirmanus M N 2004 *Rev. Sci. Instruments* **75** 1081
[15] Ma Y W, Awaji S, Watanabe K, Matsumoto M and Kobayashi N 2000 *Solid State Commun.* **113** 671
[16] Ishikawa F, Koyama K, Watanabe K and Wada H 2003 *Jpn. J. Appl. Phys.* **42** L918
[17] Koyama K, Watanabe K, Kanomata T, Kainuma R, Oikawa K and Ishida K 2006 *Appl. Phys. Lett.* **88** 132505

Development of an extraction type magnetometer under low temperature and high magnetic fields over 20 T by the hybrid magnet

K Koyama, R Sakakura and K Watanabe

High Field Laboratory for Superconducting materials, Institute for Materials Research, Tohoku University, Sendai 980-8577, Japan

E-mail: kkoyama@imr.tohoku.ac.jp

Abstract. An extraction-type magnetometer has been developed, which is performed under the low temperature of 0.5-0.6 K using a ^{3}He-refrigerator and high magnetic fields up to 18 T using a superconducting magnet (SM) and 27 T using a hybrid magnet (HM). Magnetization curves can be measured with the absolute value over 0.0005 emu using SM and 0.005 emu using HM. We confirmed that the resolution is 0.001 emu for SM and 0.005 emu for HM. For demonstrating the ability of the magnetometer, high field magnetization curves of $NdCu_4Ag$ in 0.6-4.2 K are presented.

1. Introduction

Magnetization is one of the most important properties for investigation of magnetic materials. Especially, the measurements under multi-extreme conditions such as high magnetic field, high pressure and low temperature provide us useful information on the intrinsic properties. Therefore, many research groups have developed magnetometers that are utilized under multi-extreme conditions and showed interesting results so far [1-6]. At the High Field Laboratory for Superconducting Materials (HFLSM), high-field magnetization curves have been measured by a sample extraction magnetometer combined with hybrid magnets (HM). To avoid the influence of large leakage magnetic field from HM, the sample was extracted by an air-piston technique. However it is difficult to precisely control the sample motion using this technique. On the other hand, HFLSM also developed a vibrating sample magnetometer system using a nonmagnetic ultrasonic motor, which is a small system and can be equipped on HM [7]. The result of the development indicated that the non-magnetic ultrasonic motor is very useful for precisely controlling the sample position and motion in the large leakage of the magnetic field generated by HM. Recently, HFLSM developed another extraction type magnetometer using the ultrasonic motor. The magnetometer has been performed under pressures up to 1.2 GPa using a miniature high-pressure cell in magnetic fields up to 27 T and in the temperature range of 1.7-300 K [8,9].

In this work, we have developed a new extraction-type magnetometer so that the magnetization measurement can be carried out under temperature T of 0.5 K and high magnetic fields up to 27 T. Obtained results show that magnetization curves can be measured with the absolute value over 0.0005 emu using a superconducting magnet (SM) and 0.005 emu using HM. In addition, we confirmed that the resolution is approximately 0.001 emu for SM and 0.005 emu for 28T-HM. For demonstrating the ability of the instrument, we will present results of the high-field magnetization curves for $NdCu_4Ag$ in 0.6-4.2 K under magnetic fields up to 27 T using 28T-HM.

2. Construction of the magnetometer

In Fig.1, we show the schematic diagrams of the developed magnetometer in magnetic fields up to 28 T produced by 28T-HM and in the temperature range from 0.5 to 4.2 K. This magnetometer mainly consists of an extraction system driven with the nonmagnetic ultrasonic motor, pickup-coils, a ^{3}He refrigerator, a personal computer (PC) and the hybrid magnet (a water-cooled magnet WM and a superconducting magnet (SM). This system is designed on the basis of the magnetometer reported by Koyama *et al.* [2]. The refrigerator consists of a ^{4}He-cryostat with a ^{4}He gas handling system and a ^{3}He insert with a ^{3}He gas handling system. The ^{4}He insert is utilized for the condensation of the ^{3}He gas. The sample holder (Delrin) including the sample is cooled directly with liquid ^{3}He. Removing the ^{3}He insert and ^{3}He gas-handling system, we can also measure the magnetization in a wide temperature range 1.7 K $\leq T \leq$ 300 K. In our system, the ultrasonic motor (FUKOKU) controls the sample motion and position precisely by a built-in optical encoder. The sample set in a sample holder is extracted in the pair of pickup coils using the timing belt that connects the ultrasonic motor. The motor can lift a loading weight of 600 g and control the sample position with acuracy of ± 0.1 mm. The maximum speed for the extraction is 70 mm/s. The extraction and the sample position are controlled by PC.

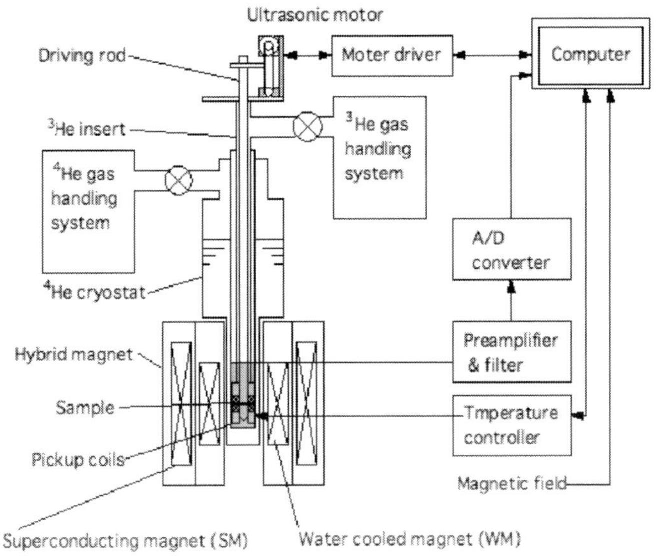

Figure 1. Schematic diagram of the developed magnetometer which is installed in a hybrid magnet consisting of a water cooled magnet and a superconducting magnet.

Figure 2 shows the cross sectional view of axial pickup coils for detecting the magnetization, the sample holder with a driving rod, and the lower port of the ^{4}He-cryostat and the ^{3}He-cryostat. The inner diameters of the ^{4}He cryostat and ^{3}He insert are 44 mm and 26 mm, respectively. The outer diameter of the sample holder is 6.0 mm. The pickup coils are connected in series opposition for compensating a noise signal arising from the fluctuation of the applied magnetic field and located at the center of the magnet. The coil was made by winding an insulated copper wire of 0.06 mm diameter about 4900 turns on a polycarbonate coil bobbin with the inner diameter of 7.0 mm. The width of each coil is 7.0 mm and the distance between the coil centers is 13.0 mm. In order to obtain the best compensation, the number of turns was carefully adjusted in a uniform ac magnetic field of ~100 Hz. The lower part of the driving rod is connected to the sample holder. A thermometer (Cernox) is at the center of the coils. The thermometer was calibrated in magnetic fields at 0.5 K. For the magnetization measurement, the sample holder including the sample is linearly moved up and down through the pickup coils by the driving rod connected to the timing belt. Usually, the speed of the motion is 55.4 mm/s and the sample is shifted at the distance of 15 mm. The bottom of the magnetometer is fixed using Al foil and a Cu-shield tube in the cryostats.

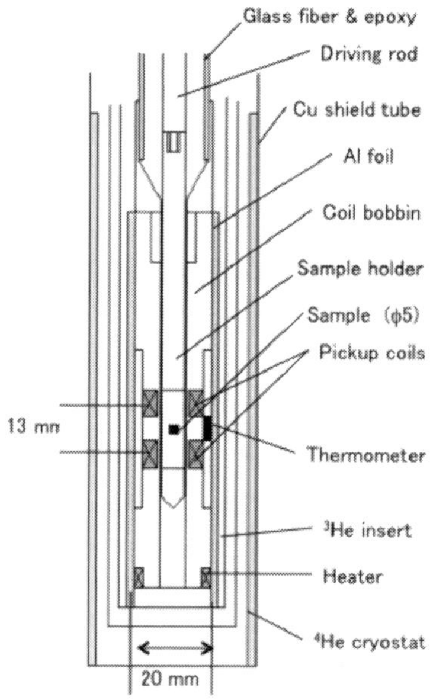

Figure 2. Cross sectional view of axial pickup coils for detecting the magnetization, the high-pressure cell assembly with a driving rod, and the lower port of the ^{4}He-cryostat.

The voltage signal induced in the pickup coils by the movement of the sample is amplified by 10-10,000 times using a low noise pre-amplifier (DL INSTRUMENTS) that has a band pass filter. The amplified signal is directly acquired in the memories of the PC though a 16 bit analog-to-digital (A/D) converter (TEAC). The magnetization is calculated from the integration of an output voltage by the computer. The calculation method for magnetization was described in the previous paper [2]. The magnetic field is taken to the PC by GPIB and RS-232C standards.

3. Performance test

3.1. Magnetization measurement of Pd

In order to calibrate the pickup coil, the magnetization of a standard Ni sample (60.92 mg) was measured at 4.2 K using a superconducting magnet. In addition, we measured the magnetization of Pd (271.9 mg) at 4.2 K using the superconducting magnet (18T-SM) in order to estimate the ability of the magnetometer. Figure 3 shows the magnetization curve of Pd at 4.2 K in field range of 0 -18 T (a), 0 - 0.3 T (b) and 17-18T (c). As seen in these figures, the magnetization can be measured for the absolute value over 0.0005 emu and the resolution is 0.0003 emu using the superconducting magnet (SM).

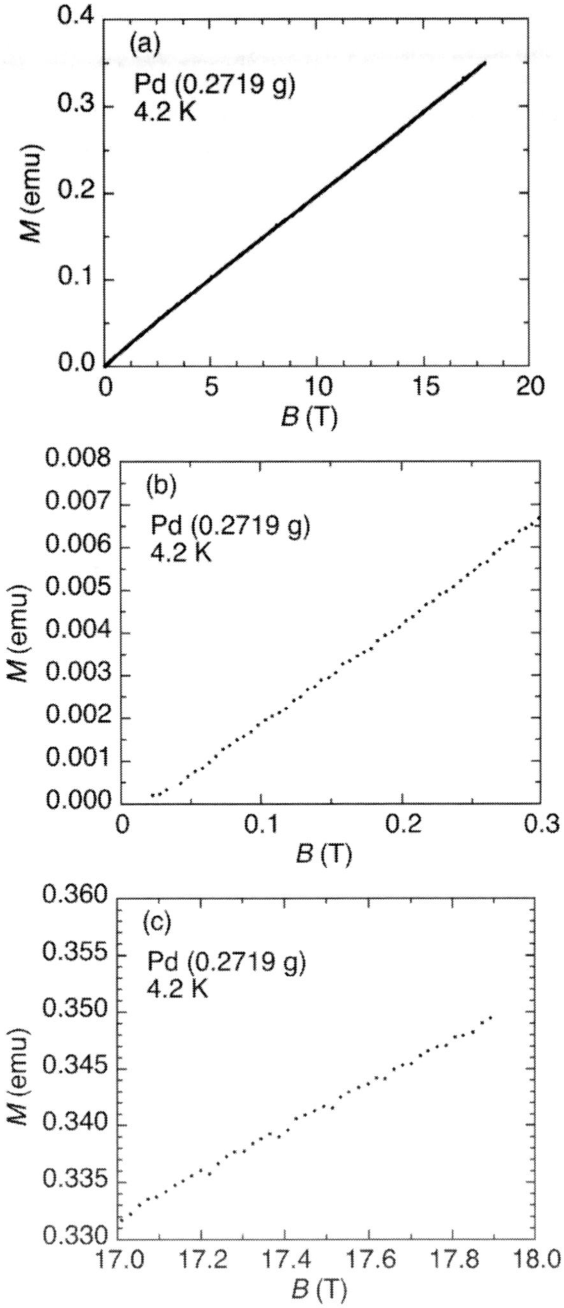

Figure 3. Magnetization curves of Pd in field range of 0 -18 T (a), 0 - 0.3 T (b) and (c) 17-18T (c) at 4.2 K.

3.2. Magnetization measurements of $NdCu_4Ag$

Intermetallic compound $NdCu_4Ag$ with the cubic $MgCu_4Sn$-type crystal structure is an antiferromagnet [10]. Recently, Adachi and Koyama reported that the Néel temperature T_N is 4.3 K using the single crystal of $NdCu_4Ag$ [11]. Their measurements show that the magnetization was not saturated even a magnetic field of 18 T at 2 K. In this work, we carried out high-field magnetization measurements of a $NdCu_4Ag$ single crystal with the weight of 10.35 mg in magnetic fields up to 27 T to examine the ability of the present apparatus. We show only some of the data to demonstrate the ability of the magnetometer. Details of the experimental results and discussions will be presented elsewhere.

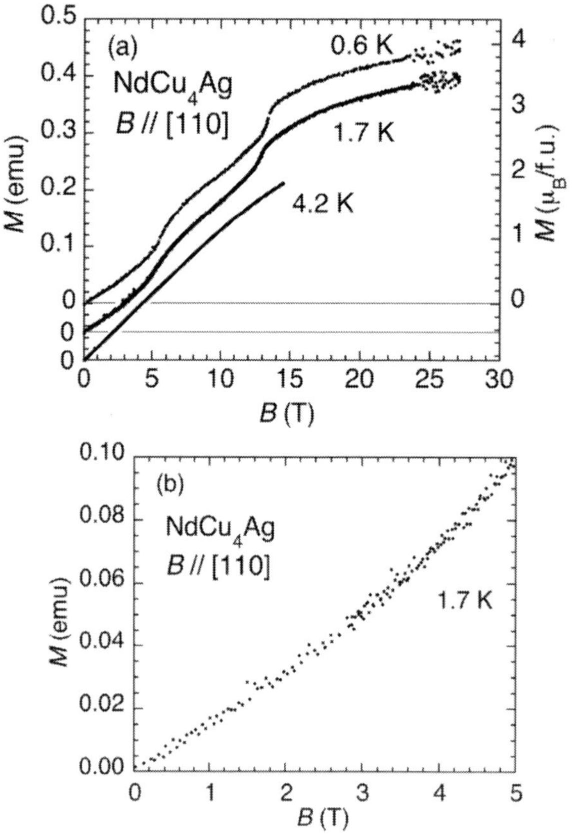

Figure 4. High-field magnetization curves of the single crystal of $NdCu_4Ag$ at 4.2 K, 1.7 K and 0.6 K in fields up to 27 T (a) and 15 T (b) using 28T-HM (a).

Figure 4(a) shows the high-field magnetization curves of the single crystal of $NdCu_4Ag$ at 4.2 K, 1.7 K and 0.6 K using 28T-HM. The single crystal was prepared by Kitai [12]. For measuring the magnetization at 4.2 K, we utilized only WM in magnetic fields up to 15 T. The temperature decreases down to 0.6 K for 28T-HM though it reaches to 0.5 K using 15T-SM. The sample-extraction rate was controlled for keeping the temperature 0.6 K.

At 4.2 K, the magnetization curve shows a paramagnetic behavior and we cannot see the field-induced phase transition in $B < 15$ T. On the other hand, we clearly observed metamagnetic transitions in the vicinity of 5 T and 13 T at 1.7 K, which increases with decreasing temperature. At 0.6 K, the metamagnetic transitions occur in the vicinity of 6 T and 14 T. The magnetic moment increases with increasing a magnetic field over 15 T, which is not saturated even in 27 T. Figure 4(b) shows the magnetization curves of $NdCu_4Ag$ at 1.7 K in $0 \leq B \leq 5$ T using HM. Figs 4 (a) and 4(b) indicate that magnetization is measured with the absolute value over 0.005 emu and the resolution is 0.005 emu in $B < 24$ T using present HM.

It should noted that any big noise due to a ripple field of WM is not seen even in a maximum field of 15 T, as seen in Fig. 4(a). However, when we generated WM under the HM mode, the data are scattered in magnetic fields over 24 T. This is due to that the magnetometer detects noise field from WM of 28T-HM. Therefore we believe that high-resolution data is obtained if the present magnetometer is utilized in high-quality magnetic field even using HM. Consequently, we succeeded in the development of high-sensitive magnetometer that is utilized under low temperature of 0.6 K and high magnetic fields over 20 T.

4. Summary

We developed the extraction-type magnetometer that is performed under the temperature of 0.5-0.6 K using the ^{3}He-refrigerator, and high magnetic fields up to 18 T using the superconducting magnet (SM) and 27 T using the hybrid magnet (HM). Magnetization curves can be detected with the absolute value over 0.0005 emu using SM and 0.005 emu using HM. We confirmed that the resolution is 0.001 emu for 18T-SM and 0.005 emu for 28T-HM. Using this magnetometer, we observed the metamagnetic transitions of $NdCu_4Ag$ in 0.6-1.7 K.

Acknowledgments

This work was performed at the High Field Laboratory for Superconducting Materials, Institute for Materials Research, Tohoku University. The authors are grateful to Prof. Y. Adachi, T. Kaneko and Prof. T. Kitai for preparing the single crystals of $NdCu_4Ag$.

References

[1] Guertin R P and Foner S 1974 *Rev. Sci. Instrum.* **45** 863
[2] Koyama K, Hane S, Kamishima K and Goto K 1998 *Rev. Sci. Inst.* **69** 3009
[3] Maki K, Kaneko, T Hiroyoshi H and Kamigaki K 1998 *J. Magn. Magn. Mater* **177-181** 1361
[4] Hamamoto T, Kindo K, Muramatsu T, Kobayashi T C, Uwatako Y, Araki S, Settai R and Onuki Y 1999 *Proc. the Phys Phenomena at High Magnetic Fields-III* (Singapore: World Scientific) p549
[5] Endo S, Yamada J, Imada S, Ishizuka M, Miyamaot S and Ono F 1999 *Rev. Sci. Inst.* **70** 2445
[6] Mushnikov N V and Goto T 1999 *J. Phys.: Condence Matter* **11** 8095
[7] Awaji S, Watanabe K and Motokawa M 1998 *Proc. 15th Int. Conf. Magnet Technology* (Beijin:Science Press) p1473
[8] Koyama K, Miura S, Okada H and Watanabe K 2006 *J. Phys.: Conf. Ser.* **51** 569
[9] Koyama K, Onodera H and Watanabe K 2008 *Physica B* **403** 1607
[10] Takeshita T, Malik S K, El-Attar A A and Wallace W E 1976 *AIP conf. Proc.* **34** 230
[11] Adachi Y and Koyama K 2007 *Annual Report of High Field Laboratory for Superconducting Materials, Institute for Materials Research, Tohoku University* ed N Kobayashi (Sendai) pp86-87
[12] Kitai T, Abe S, Yoshida H., Kawae T, Kawazoe Y and Kaneko T 2007 *J. Magn. Magn. Mater.* **310** e583

Precise measurements of diamagnetic susceptibility of benzophenone and paraffin by using a magnetic levitation technique

K Takahashi, I Mogi, S Awaji and K Watanabe

High Field Laboratory for Superconducting Materials, Institute of Materials Research, Tohoku University, Sendai, 980-8577, Japan

E-mail: kohki@imr.tohoku.ac.jp

Abstract. Measurements for temperature dependence of diamagnetic susceptibility were performed under the magnetic levitation condition. The magnetic susceptibility of a single crystal of benzophenone showed monotonous decrease toward to the melting point with increasing temperature. The minimum change of the susceptibility was detected by 1.4×10^{-12} m^3/kg. On the contrary, slight increase was observed below the melting point in the case of paraffin. The susceptibility of a paraffin melt was found to be smaller than that of the solid state. It was demonstrated that the magnetic levitation enables sensitive and contactless measurements of the diamagnetic susceptibility across the melting point.

1. Introduction

Diamagnetic materials in a magnetic field gradient receive a repulsive force. When the upward magnetic force counterbalances with the downward gravitational force, the materials can be levitated [1]. In this magnetic levitation state, such a counterbalance holds for each molecule or atom constituting the materials, hence the magnetic levitation is considered to be a quasi-microgravity condition. The magnetic levitation enables containerless melting of materials, which is a novel technique for materials synthesis using high magnetic fields [2-4]. For melting and solidifying a material without any contacts through the whole process, it is very important to know the temperature dependence of diamagnetic susceptibility of the material. It is generally known that the temperature dependence of diamagnetic susceptibility is very small or negligible. A slight change in the susceptibility, however, affects the sample position in the magnetic levitation. Actually, it was observed that a sample moves slightly upward or downward during heating process in our previous experiments. This shows that the magnetic levitation condition is quite sensitive to the magnetic susceptibility change. The magnetic levitation enables to measure the temperature dependence of diamagnetic susceptibility precisely.

For a diamagnetic material placed in the vertical bore of a solenoidal electromagnet, the magnetic force per unit mass of the material in the vertical direction F_{mag} is given by

$$F_{mag} = (1/\mu_0)\, \chi_g\, B(\partial B/\partial z), \qquad (1)$$

Journal of Physics: Conference Series **156** (2009) 012020 doi:10.1088/1742-6596/156/1/012020

where χ_g is the magnetic susceptibility per unit mass, μ_0 the vacuum permeability and B the magnetic flux density at distance z from the magnet center. When F_{mag} acts upward and balances with the downward force due to the gravity g i.e. $F_{mag} = g$, the material levitates; that is the magnetic levitation condition. In such a condition, the magnetic susceptibility of the material is evaluated as,

$$\chi_g = g\,\mu_0/B(\partial B/\partial z). \tag{2}$$

The magnetic force field $B(\partial B/\partial z)$ can be obtained from calculations of the field distribution of a magnet in which the sample levitates.

Measurements of diamagnetic susceptibility have been performed by using the magnetic levitation technique by Tanimoto *et al* [5]. They determined the susceptibilities of some plastics from $B(\partial B/\partial z)$ at different levitating positions depending on the susceptibilities at room temperature.

A sample position will change when its susceptibility decreases or increases in a heating process. When the change of the susceptibility is large, a sample will fly away or fall down by losing its stable position. In order to perform wide range measurements of the susceptibility change, it is better to control a sample position when a sample moves. Thus we demonstrate sensitive measurements of diamagnetic susceptibility of a levitating sample with adjusting the magnetic field to keep a sample in a given position.

2. Experimental

Benzophenone crystal (molecular formula: $(C_6H_5)_2CO$) belongs to the orthorhombic symmetry and the unit cell dimensions are a = 10.26, b = 12.09, c = 7.88 Å [6]. The magnetic susceptibilities of the crystal have been reported to be $\chi_a = -6.07 \times 10^{-9}$, $\chi_b = -6.11 \times 10^{-9}$ and $\chi_c = -10.3 \times 10^{-9}$ m^3/kg [7]. A single crystal of benzophenone (Wako Pure Chemical Industries, Ltd., 98.0%) with the nominal meting point of 48-50°C was employed for this work and its size was about $1.5 \times 1.5 \times 3$ mm^3.

Paraffin (C_nH_{2n+2}) is composed of single C–C bonds combining like a chain. A single C–C bond has a diamagnetic anisotropy of $\Delta\chi = \chi_{\parallel} - \chi_{\perp} = -16 \times 10^{-12}$ m^3/mol [8], where $\chi_{\parallel}$ is the magnetic susceptibility parallel to the bond and $\chi_{\perp}$ perpendicular to the bond. Paraffin with the melting point 48-56 and 56-58°C (Wako Pure Chemical Industries, Ltd.) were used for measurements. Samples for levitation experiment were cut from purchased bulk into about $4 \times 4 \times 4$ mm^3 pieces.

In order to achieve the magnetic levitation of benzophenone or paraffin, a hybrid magnet (28T-HM) was used, which consists of an inner water-cooled resistive magnet and an outer superconducting magnet. This magnet can generate up to -3200 T^2/m of $B(\partial B/\partial z)$ at the central field of 27 T. Temperature was controlled from 20 to 60°C by using a water circulating thermostatic bath set in a 52 mm room temperature bore of the magnet as illustrated in figure 1. Temperature in the bath was measured by a copper-constantan thermocouple. A sample was put in a sample cell with nitrogen gas and the cell was shield with Sealon film (Fuji Photo Film Co., Ltd.). Buoyancy caused by paramagnetic oxygen gas in the air, so called the magneto-Archimedes effect, did not affect the levitation condition in this case. The behavior of a levitating sample was observed by a CCD camera through a prism and images were recorded on a DVD recorder.

The measurement procedure was as follows. As a first step, a sample was levitated at a certain position by adjusting the magnetic field. Then temperature was raised to the next measurement temperature at a rate of less than 1 °C/min. If the sample moved upward or downward during the process, the magnetic field was adjusted in order that the sample may move back to the initial position. Figure 2 shows snap shots of a benzophenone during the process to keep a levitating position mentioned above. The value of $B(\partial B/\partial z)$ to levitate the sample at the temperature was determined after the sample kept the position stably.

3. Results and discussion

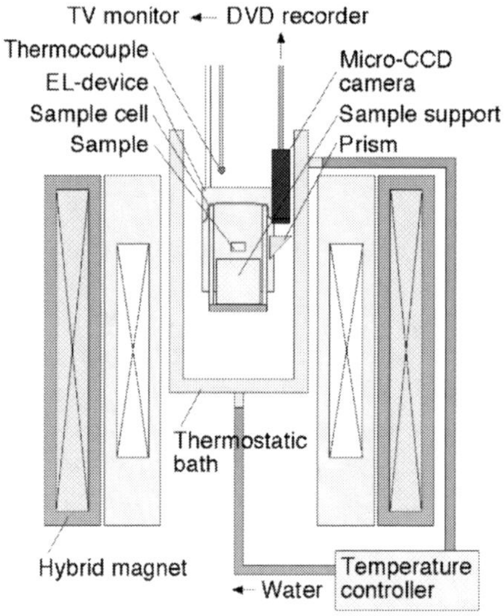

Figure 1. Schematic illustration of an experimental setup for magnetic susceptibility measurements using the magnetic levitation technique.

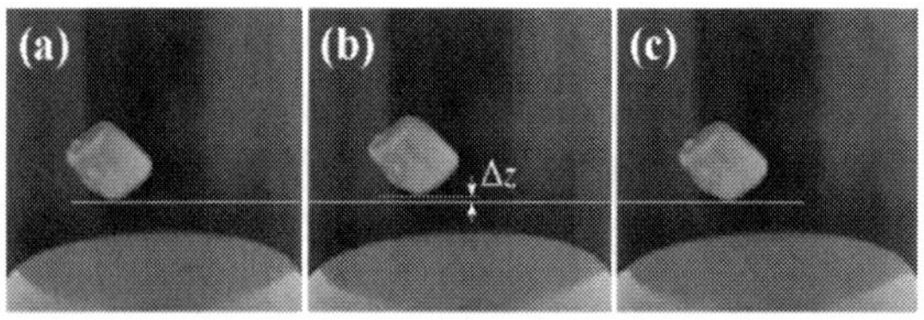

Figure 2. A position change of a levitating benzophenone during a heating process. The temperature and the magnetic field are; (a) 20°C, 22.676 T, (b) 22°C, 22.676 T and (c) 22°C, 22.675 T, respectively. Δz in this figure is about 0.15 mm.

A single crystal of benzophenone at 20°C was levitated at the central field of 22.676 T, which corresponded to $B(\partial B/\partial z)$ of −2014.0 T^2/m. The temperature dependence of the magnetic susceptibility per unit mass of benzophenone is shown in figure 3. The susceptibility decreased monotonously as increasing temperature, and decreased rapidly a few degrees below the melting point of 48°C. In the case of benzophenone, a control of the levitating position could not continue and the sample fell down because a change of the susceptibility was very large over 41°C. This temperature dependence behavior is considered as follows. At first a single crystal of benzophenone levitated as the *a*-axis is parallel to the magnetic field direction to minimize the magnetic energy, because the susceptibility along the *a*-axis is larger (smaller in the absolute value) than others. Then thermal vibration of molecules became larger with increasing temperature, and the susceptibility became smaller toward the value of the liquid state. Assuming that the value of the magnetic susceptibility in the liquid state of benzophenone is the average of values along each axis, it is to be -7.49×10^{-9} m^3/kg.

The change of the susceptibility could be detected by 1.4×10^{-12} m^3/kg as the minimum value between at 20 and 24°C in this work, it was 0.02% of the value at 20°C of -6.1189×10^{-9} m^3/kg. This indicates that the magnetic levitation is one of useful methods for sensitive measurements of the diamagnetic susceptibility change.

Paraffins were levitated at z = 98.9 mm at the central field of 18.629 T (m.p. = 56-58°C) and 18.554 T (m.p. = 48-50°C) at 20°C. Figure 4 shows the temperature dependences of the magnetic susceptibility per unit mass of paraffins. In the case of paraffins, both samples were heated to melt

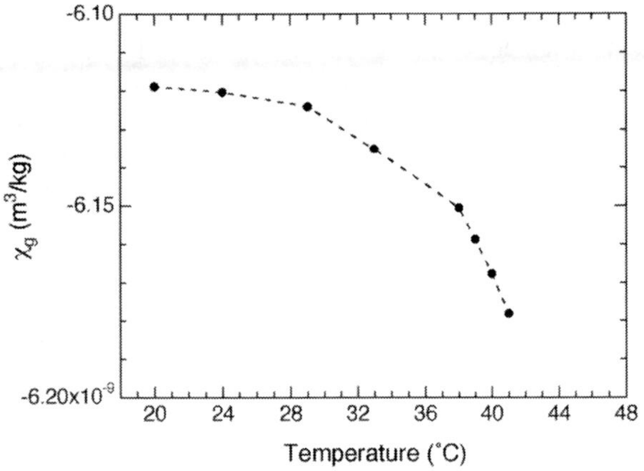

Figure 3. Temperature dependence of the magnetic susceptibility per unit mass of benzophenone.

state just above the melting point continuously. Snap shots of paraffin (m.p. = 48-50°C) in during heating process are shown in figure 5. Temperature dependence behavior of both samples were similar qualitatively and different from the case of benzophenone. The susceptibility increased with increasing temperature below the melting point and then decreased in the melt state. Paraffins used in this work were mixture of *n*-alkanes, which undergo solid-solid phase transitions just below the melting point and transition temperatures depend on the carbon number [9]. Slight increase observed in the

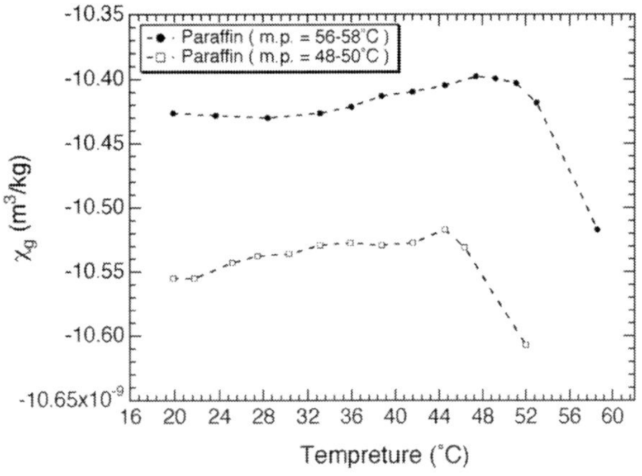

Figure 4. Temperature dependence of magnetic susceptibility per unit mass of paraffin (m.p. = 48-50 and 56-58°C).

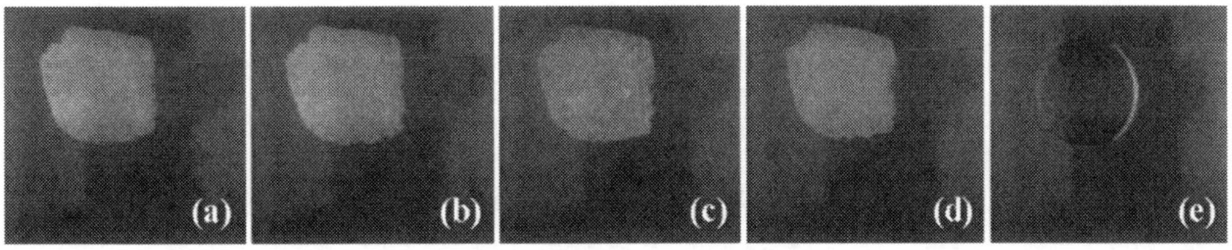

Figure 5. Snap shots of a levitating paraffin (m.p. = 48-50°C) in a heating process. Temperatures are; (a) 20°C, (b) 30°C, (c) 42°C, (d) 46°C and (e) 52°C, respectively.

temperature dependence of the magnetic susceptibility was possibly caused by this phase transitions. Increasing in the susceptibility means that the magnetic orientation may occur accompanied with the phase transitions. To confirm this, it is necessary to measure the temperature dependence of the susceptibility of a monodisperse *n*-alkane.

Magnetic susceptibility measurements by using magnetic levitation technique have advantages compared to conventional methods as follows. (1) Large size sample is not necessary for measurements. A small sample less than 1 mm in length is enough to be measured if a sample is observable on a TV monitor. (2) No sample holder is needed to support a sample in measurements. In most of cases in magnetization measurements, magnetic moment of a sample holder is to be a background to the signal of a sample. (3) In heating process, the magnetic susceptibility can be measured across the melting point continuously with *in-situ* observations as shown in figure 5. In order to obtain higher accuracy in the absolute value, the more exact value of $B(\partial B/\partial z)$ at a sample levitating position is required. However, relative change in the temperature dependence of the susceptibility can be detected sensitively by using this method as reported in this paper.

4. Conclusion

Diamagnetic susceptibility measurements of benzophenone and paraffin were demonstrated by using a magnetic levitation technique. Temperature dependences of the susceptibility were obtained for both samples. The minimum change in the susceptibility could be detected by 1.4×10^{-12} m^3/kg in the heating process of benzophenone. From the results, it was found that the magnetic levitation is one of the most useful methods for sensitive measurements of the diamagnetic materials. By using this method, temperature dependence of the susceptibility across the melting point can be obtained without any contacts.

Acknowledgments

This work was partly supported by Grants-in-Aid for Young Scientists (B), 19760043, and Grants-in-Aid for Scientific Research (C), 18560044, from the Ministry of Education, Science and Technology, Japan.

References

[1] Beaugnon E and Tournier R 1991 *Nature* **349** 749
[2] Kitamura N, Makihara M, Hamai M, Sato T, Mogi I, Awaji S, Watanabe K and Motokawa M 2000 *Jpn. J. Appl. Phys.* **39** L324
[3] Takahashi K, Umeki C, Mogi I, Koyama K, Awaji S, Motokawa M and Watanabe K 2004 *Physica* B **346-347** 277
[4] Takahashi K, Mogi I, Onogi T, Awaji S, Motokawa M and Watanabe K 2006 *Sci. Tech. Adv. Mater.* **7** 346
[5] Tanimoto Y, Fujiwara M, Sueda M, Inoue K and Akita M 2005 *Jpn. J. Appl. Phys.* **44** 6801
[6] Wyckoff R W G 1971 *Crystal Structures* 2nd ed vol 6 part 2 (New York: Interscience Publishers) chapter XVB p 32
[7] Krishnan K S, Guha B C and Banerjee S 1933 *Philos. Trans. R. Soc. London* **A231** 235
[8] Herlach F (Ed.) 1985 Topics in Applied Physics Vol 57 (Berlin: Springer) p 147
[9] Sirota E B, King H E, Singer D M and Shao H H 1992 *J. Chem. Phys.* **98** 5809

Journal of Physics: Conference Series **156** (2009) 012021 doi:10.1088/1742-6596/156/1/012021

Diamagnetic Susceptibility of Single Micro-Particles Detected by Free Translational Motions in Field Gradient

K Hisayoshi, S Kano and C Uyeda

Institute of Earth and Space Science, Graduate School of Science, Osaka University, Osaka, Japan

E-mail: hisayoshi@ess.sci.osaka-u.ac.jp

Abstract. Free translational motion of a solid particle from magnetic-field centre, caused by a repulsive force due to field gradient dB/dx, was observed in microgravity μG with negligible effect of viscous drag. The translation was visually obtained for Rochelle salt and silicon carbide and corundum. Acceleration a was obtained for corundum by use of a high vision camera, and it was experimentally confirmed that diamagnetic susceptibility χ_{DIA} per unit mass of the particle was described as proportional to a. By using this method, χ_{DIA} of small particle becomes measurable since the new method does not require a sample holder.

1. Introduction

Necessity of detecting magnetic properties of a single particle has increased recently, in connection with the growing interest on nano-sized materials [1]. Diamagnetic susceptibility χ_{DIA} of small particles were detected recently by introducing large field gradient [1, 2]. Magnetic levitation of diamagnetic materials caused by field-gradient force was realized in strong field above 10 Tesla [3-6]. The above-mentioned effects were all observed in normal gravity in a presence of viscous drag. Field gradient force has been conventionally used in separating particles that possess ferromagnetic or paramagnetic susceptibility.

Free translational motion has been recently observed for a diamagnetic sample released in microgravity μG [7]. The sample was ejected from field center the motion in a course of observing field-induced rotational-oscillation of mm-sized sample; diamagnetic anisotropy $\Delta\chi_{DIA}$ of mm-sized sample was obtained with high sensitivity from period of the above oscillation. It is expected that χ_{DIA} of a small particle can be obtained from acceleration a of the above translation, provided that spatial field-distribution is known. In the present report, possibility of detecting χ_{DIA} on micron-sized particles is discussed based on observed results on mm-sized samples. χ_{DIA} and $\Delta\chi_{DIA}$ values are not obtained as yet for most of the inorganic materials. It is important to compile these values, since they are basic parameters that characterize various magnetic field effects that are recently reported on diamagnetic materials [8].

2. Experimental

Motion of translation from field center was observed for mm-sized crystals of Rochelle salt, silicon carbide and corundum that were released in microgravity. An apparatus that was newly developed to observe translation is shown in Fig.1, which is based on a system previously developed to measure magnetic anisotropy [7]. The system was composed of a permanent dipole magnet (NEOMAX X-

1466) which can produce a homogeneous field of $\boldsymbol{B}$ =1.18T in a spherical region of $1.5\text{x}10^{-2}$ m in diameter at center of pole pieces. Radial direction of dipole is defined hereafter as x-axis, whereas an axial direction (direction between the poles) is defined as z-axis; field centre is identical to origin. Spatial distribution of field intensity along x-axis was measured prior to transition experiment. Field intensity reduces monotonously along $\pm x$-axis.

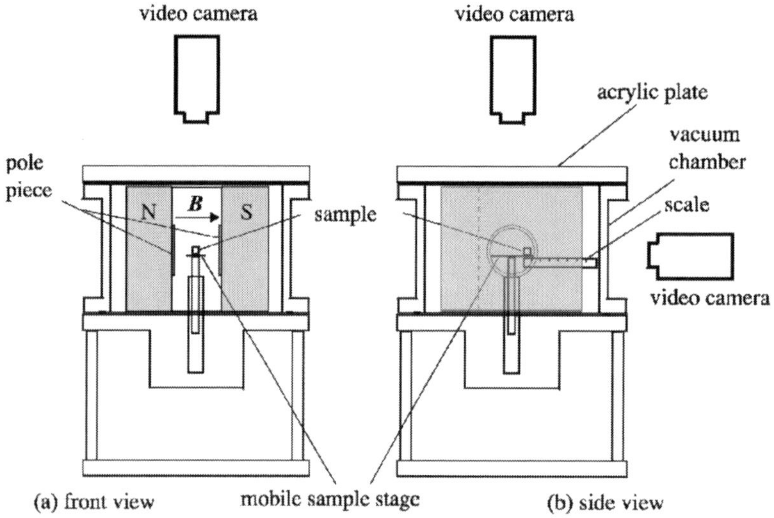

Figure 1. Schematic view of apparatus developed to observe motion of sample caused by field gradient force. Left image (a) is front view and right image (b) is side view.

Crystal samples were placed on a mobile sample stage, at a position just out side above-mentioned homogeneous field along x-axis (x =1.73cm) as described in figure 1. Experiment was performed in a reduced pressure of 10 Pa, in order to exclude viscous drag of the gas. The apparatus was loaded on a drop capsule at a μG facility; the μG experiments were performed at two facilities, namely at Micro Gravity Laboratory of Japan (MGLAB, Toki, Gifu, Japan) and at National Institute of Advanced Industrial Science and Technology, AIST, (Sapporo, Hokkaido, Japan). Duration of μG at MGLAB was 4.5 s with remaining gravitational acceleration of $2.0\text{x}10^{-3}$ Gal [9]. At AIST, the duration was 1.3s with remaining gravitational acceleration of $5.0\text{x}10^{-1}$ Gal [10]. The stage was removed from its initial position with high velocity (~ 0.4 m/s) shortly after achievement of microgravity (~ $5\text{x}10^{-3}$ s) following a procedure used in previous studies [7, 11]; this procedure was effective to minimize initial velocity of sample. Measured samples were synthetic single-crystals and their sizes were about $2\text{x}2\text{x}2\text{mm}^3$. Concentration of magnetic ion was below 1ppm for all samples according to magnetic and chemical analyses. Increase of χ value caused by paramagnetic susceptibility is below 10^{-8} emu/g when 1ppm of magnetic ion is contained in a diamagnetic oxide. The translational motion of sample was visually observed using a video camera from a direction of x-axis and z-axis. In case of corundum, value of acceleration a was analyzed with static images transformed from moving images recorded by hi-vision video camera. The spatial resolution of static images was 0.08mm and the time resolution between static images was 1/29.97 sec. The above resolution gives the upper limit of χ_{DIA} sensitivity, since χ_{DIA} is dependent on value of a obtained from the static images.

3. Results and Discussion

Examples of observed translations are shown in figure 2 for the three samples. It is seen that diamagnetic samples are accelerated by field gradient as expected. In a free space with negligibly small effects of gravity and viscosity, acceleration a is described as ma = grad (½ $m\chi_{DIA}B^2$) for diamagnetic particle of mass m. Accordingly χ_{DIA} is described as

$$a = \chi_{DIA}\left(\frac{dB}{dx}\right)B \quad (1)$$

χ_{DIA} of the particle is obtained by inserting measured values of a, B and dB/dx in the above equation.

Efficiency of the above method to obtain χ_{DIA} was quantitatively examined using numerical relationship between a and $(dB/dx)B$ observed for corundum. χ_{DIA} was obtained from gradient of a regression line of eq.(1) based on 11 sets of $[a, (dB/dx)B]$ data as described in figure 3. Numerical values of a, B and dB/dx used in the calculation are listed in Table 1. Correlation coefficient of between experimental data and regression lines using least-squares method was 9.7, which indicate that the observed translation is consistent with eq.(1) for corundum. Obtained χ value is $(-2.64 \pm 0.03) \times 10^{-7}$ emu/g which is rather small compared to a published value of corundum which is -3.63×10^{-7} emu/g; the deviation between measured and published χ_{DIA} may derive from a paramagnetic impurity grain that is accidentally attached to crystal during sample setting. In order to examine the generality of the relationship described in eq.(1), the above-mentioned experiment should be performed on many diamagnetic materials with different χ values, and in different field distributions. According to quantitative evaluations based on eq.(1) as well as on size of experimental area, it is expected that the new method is capable of detecting χ_{DIA} between 1×10^{-7} and 1×10^{-5} emu/g in microgravity with

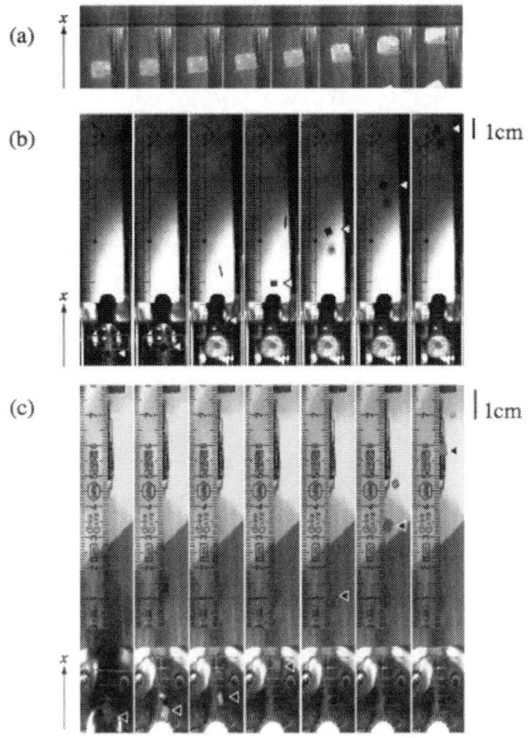

Figure 2. Visual image of translational motion of (a) Rochelle salt, (b) silicon carbide and (c) corundum. The images are arranged in the order time from left to right; time intervals between the images are (a) 0.167sec, (b) and (c) 0.334sec. Length of scale bar is 1cm in (b) and (c). Magnetic field is applied in the vertical direction in all the figures. Upper side of the images is $B = 0$. High vision images were recorded only for corundum; acceleration a is obtained from these images.

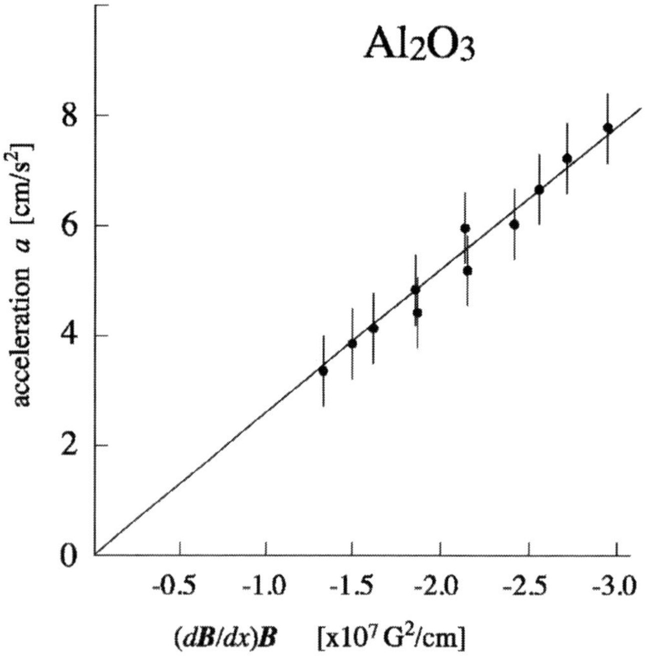

Figure 3. Relationship between acceleration *a* and $(dB/dx)B$ for corundum. Error bar of acceleration derives from measurement of position.

Table 1. Numerical data measured for corundum.

position[a] (cm)	acceleration *a* (cm /s^2)	***B*** (G)	$d\boldsymbol{B}/dx$ (G/cm)	χ_{DIA} (x10^{-7} emu/g)
2.19	3.4	11068.2	-1182	-2.63
2.22	3.9	11026.3	-1339	-2.66
2.31	4.2	10888.8	-1464	-2.64
2.35	4.5	10835.6	-1702	-2.44
2.40	4.9	10742.4	-1703	-2.69
2.46	5.3	10632.7	-2003	-2.47
2.49	6.0	10554.4	-2003	-2.86
2.57	6.1	10392.5	-2301	-2.55
2.63	6.7	10237.8	-2473	-2.66
2.71	7.3	10027.9	-2684	-2.71
2.78	7.9	9818.5	-2975	-2.69

[a] Distance from the centre of magnetic field along *x*-axis (see text).

duration of 4.5s. According to a Pascal rule, χ_{DIA} of a material is approximately equivalent to sum of diamagnetic susceptibility numerically assigned to individual atoms. An approximate χ_{DIA} value of unmeasured material is predicted from this rule; most of the calculated values are within the above-mentioned range. Sensitivity of susceptibility measurement in a conventional SQUID method is 10^{-8} emu. In contrast, this new method has sensitivity of 10^{-10} emu.

Lower limit of spatial resolution is improved to a level of an optical microscope for observing the above-mentioned translation Hence χ_{DIA} becomes measurable on micron-sized sample According to preliminary experiments in μG, a diamagnetic sample translate along a rigid pass having minimum magnetic potential. Hence motions of particle may be observed by a microscope with shallow focus depth. Field induced translational motion of micron-sized particle dispersed in fluid has been observed by an optical microscope. [1].

4. Conclusion

Detection of χ_{DIA} becomes possible on micron-sized sample by observing acceleration of free translation in field gradient as described in eq.(1). This is because the new method is free of a sample holder, and the observed motion is caused only by magnetic force; conventional magnetization measurements in normal gravity are generally prevented by background signal of sample holder when size of sample is smaller than 1mm in diameter.

References

[1] Suwa M and Watarai H 2002 *Anal. Chem* **74** 5027-5032
[2] Tanimoto Y, Fujiwara M, Sueda M, Inoue K and Akita M 2005 *Japan. J. Appl. Phys.* **44** 6801-6803
[3] Brant E H 1989 *Science* **243** 349-355
[4] Beaugnon E and Tournier R 1991 *Nature* **349** 470
[5] Berry M V and Geim A K 1997 *Eur.J.Phys.* **18** 307-313
[6] Kitamura N, Makihara M, Sato T, Hamai M, Mogi I, Awaji S, Watanabe K, Motokawa M 2001 *J. Non-Crystalline Solids* **293-295** 624-629
[7] Uyeda C, Mamiya M, Takashima R, Abe T, Nagai H and Okutani T 2006 *Japan. J. Appl. Phys.* **45** L124-127
[8] Various papers appering in *Proc. Int. Symp. New Magneto-Sci. Tsukuba* 1999 Jpn. Sci. and Tech. Corp., NIMS
[9] Iwakami T 2005 *Proc. 7th Drop Tower Days* (Berlin) p 137
[10] Okutani T, Minagawa H, Nagai H, Nakata Y, Ito Y, Tsurue T and Ikezawa K 1999 *Ceram.Eng.Sci.Proc.* **20** 215
[11] Uyeda C, Tanaka K and Takashima R 2003 *Japan. J. Appl. Phys.* **42** L1226

Ab Initio Quantum Chemical Calculation as a Tool of Evaluating Diamagnetic Susceptibility of Magnetically Levitating Substances

Y Fujiwara[1] and Y Tanimoto[2]

[1] Department of Mathematical and Life Sciences, Graduate School of Science, Hiroshima University, Higashi-Hiroshima, Hiroshima 739-8526, Japan

[2] Faculty of Pharmacy, Osaka Ohtani University, Nishikiorikita, Tondabayashi 584-8540, Japan

E-mail: fuji0710@sci.hiroshima-u.ac.jp

Abstract. On magnetic force evaluation necessary for magnetically levitated diamagnetic substances, isotropic diamagnetic susceptibility estimation by the ab initio quantum chemical calculation using Gaussian03W was verified for more than 300 molecules in a viewpoint of the accuracy in the absolute value and the calculation level affording good cost performance. From comparison, the method of B3PW91 / 6-311+G(d,p) was found to give the adequate absolute value by the relation of (observed) = (1.03 ± 0.005) x (calculated) - (1.22 ± 0.60) x 10^{-6} in a unit of cm^3 mol^{-1} and good cost performance.

1. Introduction

Owing to recently readily available superconducting magnets possessing a strong magnetic field gradient, many studies have been carried out for diamagnetic substances in a magnetic levitation environment, which is created by magnetically simulated microgravity resulting from a balance between gravitational and upward diamagnetic forces simultaneously working on the substances [1-4]. The diamagnetic force is evaluated by a product of the isotropic magnetic susceptibility of diamagnetic substances, the applied magnetic field intensity, and the magnetic field gradient. However, in some cases of photo-generated labile molecular reaction intermediates and/or molecular assemblies such as J-aggregates in equilibrium with the component molecules, in general it is not easy to directly measure the susceptibility by means of SQUID (Superconducting Quantum Interference Device) due

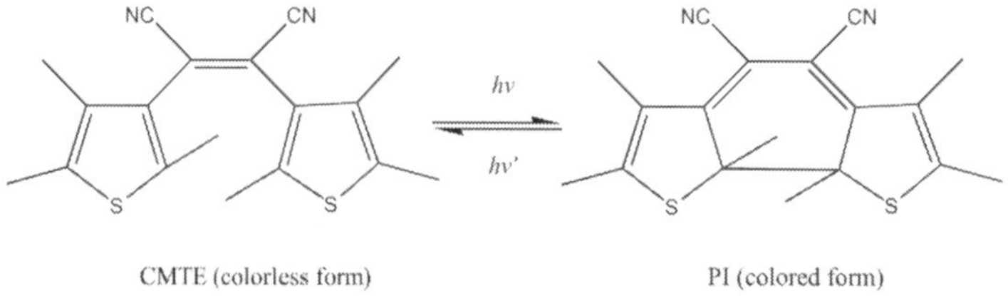

Scheme 1. A photochemical reaction scheme yielding a colored solution upon photo-irradiation (*hν*).

to the instability of intermediates and/or the mixture of substances.

Recently, Tanimoto's group elucidated intriguing effects of the magnetically simulated microgravity and hypergravity on a movement behavior of photo-chemically colored solution surrounded by an un-reacted, colorless solution inside a cuvette [5, 6]. The scheme1 illustrates one of the photoreactions upon irradiation [5].

The behavior dependent on the microgravity and hypergravity was considered by magnetic and gravitational forces operating on each diamagnetic substance (CMTE and PI as solutes and benzene as a solvent) lying before and after photo-irradiation, respectively. However, the PI solute in benzene was too unstable against the light to isolate purely and keep it stably while the susceptibility measurement with SQUID. Thus, the isotropic diamagnetic susceptibility estimation using the ab initio quantum chemical calculation was performed by a combination of a method B3LYP based on the density functional theory (DFT) and a basis set 6-31G, which was considered to be a proper calculation level in those days. However, it was not in advance confirmed whether the calculation procedure gives exact absolute values. In the other report, further, the experimental observation of efficient magnetic orientation of molecular assemblies, porphyrin-derived nanorods, caused by hypergravity was also interpreted by an imbalance in the magnetic and gravitational forces working on the diamagnetic nanorods and surrounding solvent. Yet, the accurate susceptibility estimation of the nanorod still remains difficult due to equilibrium of the rod with the precursor, porphyrin [7].

In such cases, a theoretical calculation method giving exact absolute values of the susceptibility would wield its power. Although some successful semi-empirical calculation methods such as a Pascal's method are reported so far for the purpose [8], unfortunately there seems to be no semi-empirical method which does not only afford the accurate absolute value of susceptibility but also is available throughout various chemical compounds regardless of the component chemical element. In other words, taking account of the current situation of many kinds of diamagnetic substances applied to the magnetic levitation environment, one useful theoretical calculation method capable of treating such diamagnetic molecular compounds simultaneously and consistently is indispensible.

By the way, an ab initio quantum chemical calculation program such as Gaussian is well-known to treat a wide variety of chemical elements. And, there were studies treating the susceptibility calculation for as small size of molecules as benzene by the ab initio calculation [9-12]. However, the size of target molecular compounds which we hope to know the susceptibility is much larger than benzene, as PI shown in scheme 1. In this work, therefore, the commercially purchasable ab initio program Gaussian was verified both about whether it can give the accurate absolute values of isotropic diamagnetic susceptibility in as many large compounds composed of various chemical elements as possible and about the calculation level affording good cost performance. Additionally speaking of an important reason for the verification, this calculation method provides the anisotropic terms of susceptibility at the same time [11, 12], being greatly helpful when the magnetic orientation based on the susceptibility anisotropy is associated with the magnetic force as observed in the above-mentioned nanorods under the hypergravity environment [7], though it is not referred in this study.

2. Calculation

In the isotropic diamagnetic susceptibility estimation by the ab initio quantum chemical calculation, the accuracy in absolute values and wide availability for many large molecular compounds were verified by means of a commercially purchasable calculation program of Gaussian03W (Gaussian, Inc.) for a Windows operating system. Two revisions of B.04 and E.01 of the program were used properly for a single central processing unit (CPU) and for multi-CPUs, respectively, according to hardware of an operated computer (1 - 8 pieces of Intel Xeon CPUs operated at one frequency in a range of 2.4 - 3.6 GHz). As a target of comparison in the calculation, more than 300 diamagnetic molecular compounds were chosen, whose experimental susceptibilities were already measured with instruments such as SQUID. Typical compounds whose susceptibilities are listed in the familiar literature often referred [13, 14] were mainly taken in this study.

Prior to the susceptibility estimation in each target molecular compound, the structural optimization was carried out by a frequently used combination of a B3LYP method [15, 16] based on the density functional theory (DFT) [17] and a 6-31+G(d) basis set [18], abbreviated to B3LYP / 6-31+G(d) hereafter. Subsequently, the susceptibility estimation with a IGAIM method [11] was attempted in respective combinations of the "method / basis set" of B3LYP / 6-31G, B3PW91 / 6-311+G(d), and B3PW91 / 6-311+G(d,p) [15, 19, 20], where a B3PW91 method is superior to B3LYP in the calculation accuracy. The superiority comes from a difference in the DFT procedures. The general sense on the order of superiority in the basis sets is 6-311+G(d,p), 6-311+G(d), 6-31+G(d), and 6-31G. The superiority of the first basis set to the second one arises from the enhanced contribution of a hydrogen atom. The superiority of the second one to the third one is based on the number of orbital functions utilized, and that of the third one to the last one depends on addition of diffused orbital function and the enhanced contribution of heavy atoms besides a hydrogen atom.

3. Results and Discussion

In use of this kind of theoretically calculated values in place of actually observed ones, how exactly the calculation procedure reproduces the observed values is a key point. Actually, however, the degree of reproduction greatly changes depending on the calculation procedure. In such a case, therefore, it is a general way to discuss the calculation precision by comparing many compounds in order to dig out the adequate calculation procedure. Further, it is a usual manner to induce a relationship applied to the calculated and observed values from the discussion and make a deduction about an individual value by means of its relationship. In fact, calculations for *g* values of paramagnetic species and infrared frequencies of organic compounds were carried out by this manner [21, 22]. Thus, we obey the same manner in this work and introduce an example of photochemistry shown in scheme 1 to which was applied the relationship between the calculated and observed isotropic diamagnetic susceptibility values and thereby which was discussed more quantitatively.

3.1. Calculation of Isotropic Diamagnetic Susceptibility Values

In the first place, we checked how exactly the calculation procedure of B3LYP / 6-31G, which was used in the study by Tanimoto's group [5], estimates an experimentally observed susceptibility value. However, its procedure never reproduced the experimental value even for a typical compound of benzene. The calculated value (-35.6×10^{-6} cm^3 mol^{-1}) did not reach even 70 % of the experimental one (-54.8×10^{-6} cm^3 mol^{-1}) [13, 14]. Therefore, we initiated to upgrade the calculation procedure and seek its reproducibility of the value according to a usual manner described above.

As mentioned above, the more than 300 target diamagnetic molecular compounds used for the isotropic diamagnetic susceptibility estimation by Gaussian03W were selected carefully so as to prove the accuracy and wide availability of the calculation method and thereby the validity of use, as follows. Their constituent chemical elements widely spread up to bromine in the periodic table, and cover usual inorganic and organic compounds. And, neutral, mono-ionic, or zwitterionic charge types of compounds are included when necessary. The compounds are also adopted, which possess aromatic groups like a benzene ring and/or the larger and more complex unsaturated chemical bonds. The compounds having a diamagnetic metal ion inside the molecular framework are included as well, and thus the largest molecular weight ranges to more than 500. As one more important viewpoint, a range of the diamagnetic susceptibility to be verified is taken into account so as to be beyond - 400 x 10^{-6} cm^3 mol^{-1}, where most of general diamagnetic compounds seem to be distributed.

Figure 1 shows results of the susceptibility calculation by a B3PW91 / 6-311+G(d) method of Gaussian03W. In two graphs on figure 1, calculated values in a horizontal axis are plotted against experimentally observed ones in a vertical axis. An area in the susceptibility range form 0 to 150 x 10^{-6} cm^3 mol^{-1} is extended in the right-hand side of figure 1 in addition to a graph depicting the overall data in the left-hand side. Approximately 270 data appear in the extended graph. The good linear relationship appears between the calculated and observed susceptibilities. As a result of the linear

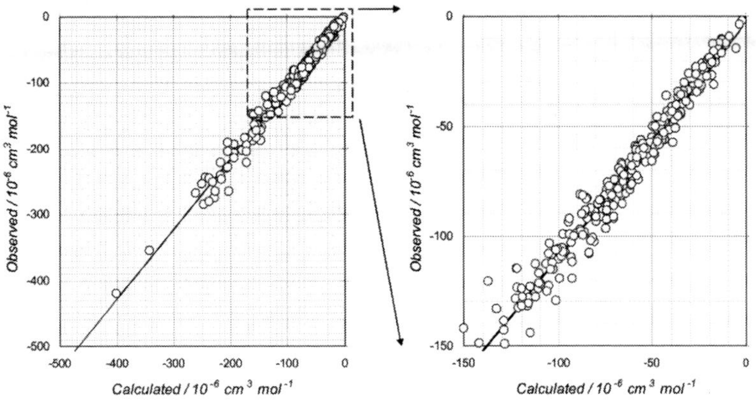

Figure 1. Plots of isotropic diamagnetic susceptibilities calculated by B3PW91 / 6-311+G(d) against observed ones.

least-squares method toward all of the data, the following fitted equation was obtained, which accordingly seems also available to estimate the susceptibility unknown due to no measurement.

(observed or unknown) = (1.06 ± 0.009) x (calculated) - $(3.17 \pm 0.87) \times 10^{-6}$ in a unit of cm^3 mol^{-1} (1)

with a correlation coefficient (0.989), indicating a good linear relationship. The two facts of the slope (1.06) very close to a unity and the negligibly small intercept (-3.17×10^{-6}) point out validity of this calculation method as a tool of directly giving fairly accurate absolute values of the susceptibility. In fact, the straight lines calculated by (1) are well superimposed on the plots in figure 1.

Next, let us see figure 2 showing the same kind of graphs in the case of a B3PW91 / 6-311+G(d,p) method where the contribution of hydrogen atoms carries weight compared with the above-mentioned case of B3PW91 / 6-311+G(d). There exists a better linear relation between the calculated and observed susceptibilities as expressed by the following fitted equation.

(observed or unknown) = (1.03 ± 0.005) x (calculated) - $(1.22 \pm 0.60) \times 10^{-6}$ in a unit of cm^3 mol^{-1} (2)

with a better correlation coefficient (0.996). Detectable effects of a 6-311+G(d,p) basis set superior to 6-311+G(d) are seen in (2). The slope (1.03) further approached a unity. The intercept (-1.22×10^{-6}) diminished toward zero more and more. Moreover, standard deviations in both values decreased. This

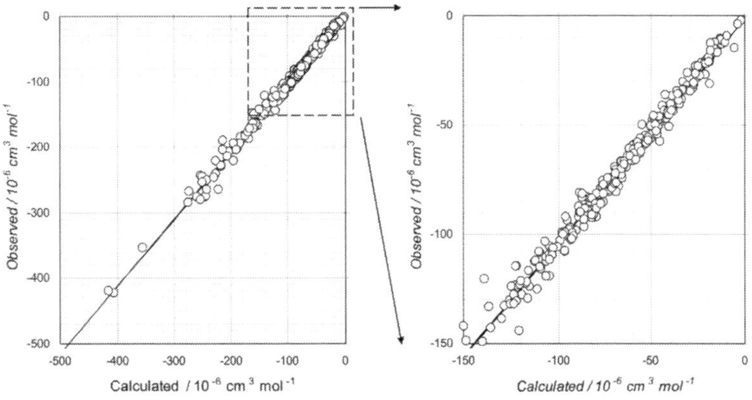

Figure 2. Plots of isotropic diamagnetic susceptibilities calculated by B3PW91 / 6-311+G(d,p) against observed ones.

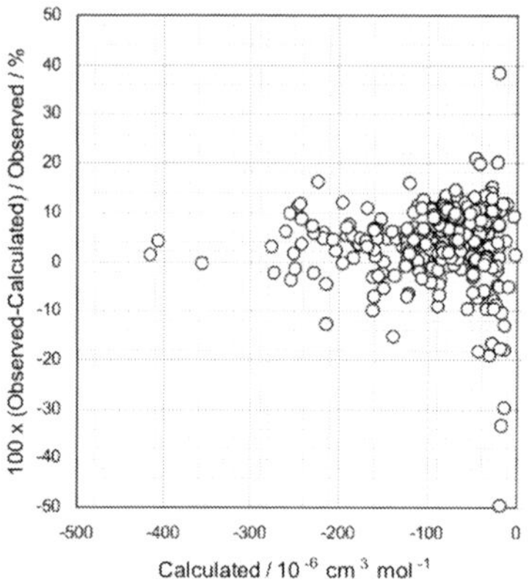

Figure 3. Degree of deviation of isotropic diamagnetic susceptibilities calculated by B3PW91 / 6-311+G(d,p) from observed ones.

effect was represented in figure 2 as a smaller extent of deviation of plots from the straight line simulated by (2) in comparison with figure 1. Figure 3 evaluates the degree of deviation with a percentage of 100 x (observed - calculated) / observed. Although there are a few plots in the noticeable divergence around -20 x 10^{-6} cm^3 mol^{-1} near zero, most (about 80 %) of plots are kept inside the value of 10 %. These results are interpreted conclusively as a result of the better basis set employed. As an inevitable consequence, therefore, (2) is a better relation formula available for the isotropic diamagnetic susceptibility estimation. The parameters so far obtained are summarized in table 1 again.

Table 1. Parameters α and β obtained by the ab initio calculation in this study, which are available for the estimation of isotropic diamagnetic susceptibility (cm^3 mol^{-1}) not obtained.

DFT method / basis set	Parameter α [a]	Parameter β [a] (cm^3 mol^{-1})
B3PW91 / 6-311+G(d)	1.06 ± 0.009	$-(3.17 \pm 0.87) \times 10^{-6}$
B3PW91 / 6-311+G(d,p)	1.03 ± 0.005	$-(1.22 \pm 0.60) \times 10^{-6}$

[a] Used in an equation of (observed or unknown value) = α x (calculated value) + β.

In the calculation here, all of the target compounds were respectively treated as those in a vacuum, meaning that they are respectively an isolated compound lacking interactions with other substrates such as themselves or solvents. As depicted in figures 1 and 2, however, the calculated values were straightforwardly compared with the observed ones which were practically measured in chemical media. This indicates that the relationships of (1) and (2) include contribution of various interactions of many compounds with the media comprehensively. Therefore, the scattering in figure 3 might partly result from such a treatment. The relatively large deviation in a small size of compounds might arise due to the same reason if they specifically tend to be affected by the surrounding conditions. Thus, the calculation should be carried out originally by the procedure capable of expressing such interactions individually. For instance, in the case of a solution phase where a reaction in scheme 1 occurs, influence of a solute - liquid solvent interaction should be taken into account. However, since no meaningful difference was detected between preliminary model calculations without and with the influence of a solvent, a direct use of the relationship of (2) was considered to be the best way at the present stage to estimate the magnetic force.

3.2. *Application of the Calculated Susceptibility Values to the Photo-Generated Labile Reaction* Intermediate

As described in the section of introduction, Tanimoto's group had already employed the ab initio calculation for the isotropic diamagnetic susceptibility estimation of solutes CMTE and PI to explain the movement of a photo-chemically colored solution [5]. The fundamental combination of B3LYP / 6-31G used was considered to be a proper calculation level in those days. The calculated susceptibilities and resulting forces are summarized in Table 2 together with the results by (2) of this study.

Table 2. Isotropic diamagnetic susceptibilities of CMTE and PI obtained by the ab initio calculation, differences in their diamagnetic forces under each of the microgravity (-1300 T^2 m^{-1}) and hypergavity (1000 T^2 m^{-1}), and their total forces expressed with a volume V of the PI solution.

Study	Isotropic diamagnetic susceptibility (10^{-6} cm^3 mol^{-1})		Difference in diamagnetic forces between CMTE and PI (10^{-5} N)		Total force (N)	
	CMTE (colorless)	PI (colored)	Microgravity	Hypergravity	Microgravity	Hypergravity
The study of Tanimoto's group[a]	-149.3481	-149.3517	7.0 V	-5.4 V	0.41 V	10.09 V
This study by (2)	-195.3434	-180.3303	-29000 V	23000 V	0.12 V	10.32 V

[a] Taken from [5].

In the study of Tanimoto's group, the total force operating on the colored solution generated upon photo-irradiation was explained by the following equation.

(total force) = (buoyancy based on the density change of solution as a difference in gravitational forces)
+ (magnetic force on the solvent benzene based on the susceptibility change by Joule heating)
+ (magnetic force based on the CMTE and PI susceptibilities)

And, it was found that a difference (0.0036 x 10^{-6} cm^3 mol^{-1}) in the calculated susceptibilities (-149.3481 x 10^{-6} and -149.3517 x 10^{-6} cm^3 mol^{-1}) of the two solutes was too tiny to recognize a meaningful difference in their diamagnetic forces (7.0 V x 10^{-5} and -5.4 V x 10^{-5} N) of the microgravity and hypergravity environments, indicating no significant responsibility of their magnetic forces on the movement of colored solution. Although the large difference in diamagnetic forces means large contribution to the total force, the value in an order of 10^{-5} N is anyway too small. Actually, the total forces (0.41 V and 10.09 V N) on the colored solution were successfully explained by the buoyancy and the magnetic force on solvent benzene originated in the susceptibility change by Joule heating upon photo-irradiation.

By contrast, it was found that the susceptibilities (-195.3434 x 10^{-6} and -180.3303 x 10^{-6} cm^3 mol^{-1}) of the two solutes calculated in this study were apparently different from each other and the order was reversed. As a result of these new susceptibilities whose gap became larger, the difference in diamagnetic forces on CMTE and PI also became large under the microgravity (from 7.0 V x 10^{-5} to -29000 V x 10^{-5} N) and hypergravity (from -5.4 V x 10^{-5} to 23000 V x 10^{-5} N) environments, respectively. The difference between their values (-29000 V x 10^{-5} and 23000 V x 10^{-5} N) thereby widened greatly. Nevertheless, these changes resulted in having little influence on the total forces (0.12 V and 10.32 V N) compared with those (0.41 V and 10.09 V N) of Tanimoto's group. Thus, according to the reliable susceptibilities calculated by the relationship of (2) whose validity of use was verified in this study, it was reconfirmed that the total force was limited by the buoyancy and the magnetic force on the solvent rather than the magnetic forces on the two solutes, as suggested by Tanimoto's group. Therefore, the ab initio calculation with B3PW91 / 6-311+G(d,p) giving the accurate absolute value of susceptibility enabled quantitative and thereby fruitful discussion based on the refined data.

4. Conclusion

On the magnetic force evaluation necessary for magnetically levitated diamagnetic substances, the ab initio quantum chemical calculation using Gaussian03W was verified in a viewpoint of the accuracy of isotropic diamagnetic susceptibility for large size of molecular compounds composed of chemical elements up to bromine in the periodic table. From comparison of the two methods of B3PW91 / 6-311+G(d) and B3PW91 / 6-311+G(d,p), the latter was found to give the considerably accurate absolute value by the relation of (observed) = (1.03 ± 0.005) x (calculated) - $(1.22 \pm 0.60) \times 10^{-6}$ in a unit of $cm^3\ mol^{-1}$. From the proportional coefficient greatly close to a unity and negligible offset in the equation, it is not too much to say that this method directly calculates a real value unnecessary to correct. Therefore, this method is also suitable for the unknown susceptibility estimation. Although, of course, we confirmed that some of other "method / basis set" calculations more complex and more sophisticated than it afforded better results in which the deviation became narrower, the method of B3PW91 / 6-311+G(d,p) used in this study was found to be stressless and give good cost performance in both a CPU time and a price of a personal computer needed for calculation. Each of almost calculations carried out in this study was enough fast to finish in several days at most.

Acknowledgment

This work was partly supported by Grants-in-Aid for Scientific Research of Area 767 No. 15085208 (Priority Area "Innovative utilization of strong magnetic fields") from MEXT of Japan and by JSPS Asian Core Program "Construction of the World Center on Electromagnetic Processing of Materials".

References

[1] Beaugnon E and Tournier R 1991 *Nature* **349** 470
[2] Berry M V and Geim A K 1997 *Eur. J. Phys.* **18** 307
[3] Ikezoe Y, Hirota N, Nakagawa J and Kitazawa K 1998 *Nature* **393** 748
[4] Mogi I, Takahashi K and Hirota N 2006 *Springer Series in MATERIALS SCIENCE 89, Magneto-Science, Magnetic Field Effects on Materials: Fundamentals and Applications*, ed M Yamaguchi and Y Tanimoto (Japan: Kodansha and Berlin: Springer-Verlag) pp 41-70
[5] Duan W, Fujiwara M and Tanimoto Y 2004 *Jpn. J. Appl. Phys.* **43** 8213
[6] Koyama F and Tanimoto Y 2006 *Mol. Phys.* **104** 1703
[7] Fujiwara Y, Shibata N, Matsumoto Y and Tanimoto Y 2007 *J. Magn. Magn. Mater.* **310**, 2859
[8] Pasacal P 1908 *Compt. Rend.* **147** 56
[9] Van Vleck J H 1932 *The Theory of Electric and Magnetic Susceptibilities* (Oxford: Univ. Press)
[10] Bhargava S and Ray N K 1979 *Indian J. Chem.* **18A** 508
[11] Keith T A and Bader R F W 1992 *Chem. Phys. Lett.* **194** 1
[12] Ruud K, Helgaker T, Bak K L, Jörgensen P and Jensen H J A 1993 *J. Chem. Phys.* **99** 3847
[13] Gupta R R 1986 *LANDÖLT-BORNSTEIN Numerical Data and Functional Relationships in Science and Technology New Series* vol 16, ed Hellwege K -H and Hellwege A M (Berlin: Springer-Verlag)
[14] 2008 *CRC Handbook of Chemistry and Physics*, ed Lide D R (Boca Raton, FL: CRC Press / Taylor and Francis)
[15] Becke A D 1993 *J. Chem. Phys.* **98** 5648
[16] Lee C, Yang W and Parr R G 1988 *Phys. Rev. B* **37** 785
[17] Hehenberg P and Kohn W 1964 *Phys. Rev.* **136** B864
[18] Ditchfield R, Hehre W J and Pople J A 1971 *J. Chem. Phys.* **54** 724

[19] Perdew J P, Chevary J A, Vosko S H, Jackson K A, Pederson M R, Singh D J and Fiolhais C 1992 *Phys. Rev. B* **46** 6671
[20] McLean A D and Chandler G. S. 1980 *J. Chem. Phys.* **72**, 5639
[21] Neese F 2001 *J. Chem. Phys.* **115** 11080.
[22] Yoshida H, Takeda K, Okamura J, Ehara A and Matsuura H 2002 *J. Phys. Chem. A* **106** 3580.

Journal of Physics: Conference Series **156** (2009) 012023 doi:10.1088/1742-6596/156/1/012023

Attempt to Detect Diamagnetic Anisotropy of Oxides with Isotropic Crystal Structure by Measuring Its Rotational Oscillation in Strong Magnetic Field

S Kano[1], K Hisayoshi[1], T Kashiwagi[2], T Kida[2], M Hagiwara[2] and C Uyeda[1]

[1]Institute of Earth and Space Science, Graduate School of Science, Osaka University, Osaka, Japan

[2]KYOKUGEN (Center for Quantum Science and Technology under Extreme Conditions), Osaka University, Osaka, Japan

E-mail: kanou@ess.sci.osaka-u.ac.jp

Abstract. Sensitivity to detect diamagnetic anisotropy $\Delta\chi_{DIA}$ of inorganic oxides was improved by increasing intensity of horizontal field from 1.6 T to 5.0 Tesla. The field induced a rotational oscillation of a magnetically stable axis of a sample, which was suspended with a thin fiber. Accurate $\Delta\chi$ values are obtained when restoration torque of the fiber is negligibly small compared to torque due to magnetic anisotropy. Accordingly, $\Delta\chi$ value at a level of 5×10^{-10} emu/g can be acquired, even if the mass of the sample is relatively small (0.318 g). Necessity of realizing $\Delta\chi$ measurement at low temperature is discussed, which realizes separation of $\Delta\chi_{DIA}$ from a paramagnetic anisotropy caused by isolated magnetic ions.

1. Introduction

Magnetic alignment of solid materials without spontaneous magnetic moment has been effectively applied in various research areas [1, 2, 3]. Intrinsic value of magnetic anisotropy $\Delta\chi$ is essential for investigating origin of magnetic alignment. Recently, small $\Delta\chi_{DIA}$ values of many popular inorganic oxides have been detected by a method using harmonic-oscillation of a sample in a horizontal magnetic field of 1.6 T [4]. This oscillation is achieved in a condition where restoration torque of a fiber suspending a sample is much small compared to torque of magnetic anisotropy. The sensitivity to detect $\Delta\chi$ in the above method is about 1×10^{-9} emu/g.

The origin of diamagnetic anisotropy was studied quantitatively based on published $\Delta\chi_{DIA}$ values recently obtained for popular oxides, namely corundum, forsterite, orthoclase, muscovite, gypsum, KDP, ADP, talc, apophirite, $Mg(OH)_2$ and $Al(OH)_3$ [5, 6]. The published values were consistently explained by assuming a constant $\Delta\chi$ tensor on individual bonding orbital composing the material [7]; here anisotropy of a single bond $\Delta\chi_{BD}$ was assumed to have a anisotropy at a level of 10^{-30}emu/g. Accordingly, $\Delta\chi_{DIA}$ of an unmeasured material can be predicted from the $\Delta\chi_{BD}$ values, and it was expected that $\Delta\chi_{DIA}$ is smaller than 1×10^{-9} emu/g for materials with high crystal symmetry. In case of wurtzite structure for example, the tetrahedral configuration in the crystal is very close to regular symmetry, so that the summation of the above-mentioned $\Delta\chi$ tensors of individual bonds in a unit cell cancels out [8]. An ideal $\Delta\chi_{DIA}$ value of a crystal with cubic symmetry is expected to be zero according to this model. Hence it is essential to improve $\Delta\chi$ sensitivity in order to understand an overall tendency on diamagnetic anisotropy of oxides.

In a previous report, $\Delta\chi$ above 1×10^{-9} emu/g were measurable even if field intensity was as small as 0.1 T, provided that a thin polyaramid fiber having a small D value was introduced [9]. It is implicated that sensitivity of $\Delta\chi$ measurement using this fiber would be considerably improved by increasing field intensity. In the present study, rotational oscillation was observed in static horizontal field of 5.0 T using a cryogen-free magnet. A new apparatus was developed for the purpose of realizing the detection of $\Delta\chi$ below 1×10^{-9} emu/g.

2. Experimental

A bulk sample is suspended in a horizontal field $\boldsymbol{B}$ by the above-mentioned polyaramid fiber that has a diameter of 1.2×10^{-5} m and a length of 3.0×10^{-2} m; the directions of two principal axes of a sample are in the horizontal plane. The direction of magnetically stable axis shows rotational harmonic-oscillation with respect to $\boldsymbol{B}$, which follows the equation described as

$$I\frac{d^2\theta}{dt^2}=-\frac{1}{2}B^2N\Delta\chi\sin2\theta-\frac{D}{l}\theta \tag{1}$$

where θ is an angle of magnetically stable axis with respect to field $\boldsymbol{B}$. I denote moment of inertia of a crystal. N is mass of a sample. D and l are tensional rigidity and length of the fiber, respectively. The evaluated D/l value of the fiber was $(1.1\pm0.4)\times10^{-9}$ Nm which was two orders of magnitude smaller compared to magnetic term in EQ.(1) [9]. Hence, the restoration torque of the fiber is negligibly small with respect to that of magnetic anisotropy in the present condition. The period of oscillation τ of a sample is deduced theoretically from eq.(1) as

$$\tau=2\pi\sqrt{\frac{I}{N\Delta\chi}}B^{-1}\left\{1+\frac{1}{4}\sin^2\frac{\theta_0}{2}+\ldots\right\}. \tag{2}$$

Here θ_0 is the angle of amplitude, which is below 15 degree in this study. $\Delta\chi$ value of the sample is determined from the gradient of measured τ^{-1}-B relationship [9].

Experimental setup developed in the present work is shown schematically in figure. 1. A cryogen-free superconducting magnet (JMTD-6T100) produced a maximum magnetic field of 5.0 T. The inhomogeniety of the field was less than 1 % in a spherical area of 10 mm in diameter at field center. The sample with fiber was put inside a glass tube at center of the homogeneous magnetic field. The rotational motion was observed through a flat mirror and recorded by a video camera (Panasonic HDC-SD3) as shown in Fig.1. A standard sample was prepared in the present experiment so as to examine the reliability of the achieved sensitivity of this system using 5.0 T. The sample, which was made from blocks of α-quartz single crystal, was expected to have $\Delta\chi_{DIA}$ value of 5.75×10^{-10} emu/g. The size of the sample was 0.50 cm square and 0.48 cm depth, and its mass was 0.318 g. Center of top plane was determined by geometrical measurement of the sample, and fiber was attached to this point. Value of I measured by rotational oscillation at B=0 condition generally agree fairly well with I value calculated from the above geometrical condition.

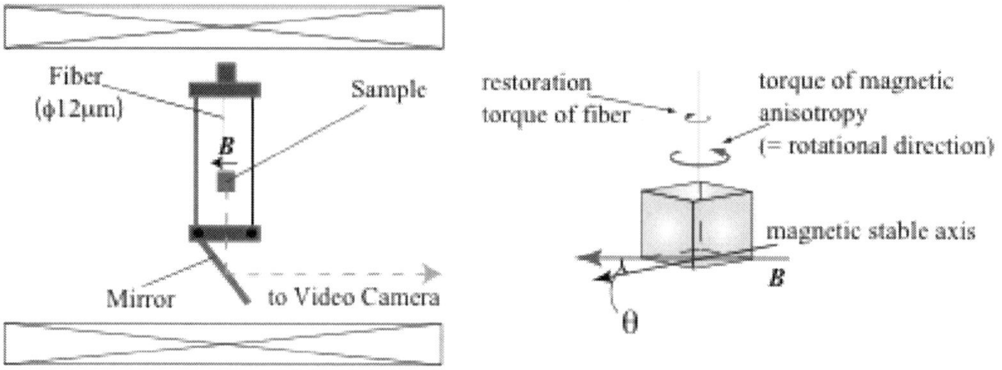

Figure 1. Schematic view of apparatus (left hand) and geometrical relations of a sample (right hand).

3. Results and Discussion

Observed τ^{-1}–B relationship of above-mentioned standard sample was shown in Fig 2. The obtained $\Delta\chi$ value was 5.2×10^{-10} emu/g, it showed good agreement with the expected value of 5.75×10^{-10} emu/g. It was confirmed that $\Delta\chi$ value at a level of 5×10^{-10} emu/g could be detected using the present setup. Origin of intercept of regression line (Fig 2.) is qualitatively explained by adding a term of viscous drag caused by air to eq.(1). Contribution of this term may be quantitatively evaluated by performing experiment in a diffused condition.

Inorganic crystals generally contain paramagnetic impurity ions that cause anisotropy of paramagnetic susceptibility $\Delta\chi_{PARA}$, and it is important to distinguish quantitative amount of $\Delta\chi_{PARA}$ in order to obtain an exact $\Delta\chi_{DIA}$. Paramagnetic susceptibility and its anisotropy are proportional to temperature T according to a Curies' law, whereas diamagnetic $\Delta\chi_{DIA}$ is independent of temperature. Previous experiments revealed that a $\Delta\chi$-T^1 relationship is effective to obtain $\Delta\chi_{DIA}$ value [6, 10, 11, 12]; here intercept of the $\Delta\chi$- T^1 regression line with the T=0 axis was adopted as a $\Delta\chi_{DIA}$ value. However when concentration of paramagnetic ion was high, the obtained $\Delta\chi_{DIA}$ value contained large amount of uncertainty. This is because deviation of measured $\Delta\chi$- T^1 data with respect to regression line was not negligible. In order to obtain a reliable $\Delta\chi_{DIA}$ value, high sensitive $\Delta\chi$- T^1 measurements should be performed on samples with high purity. By following the above-mentioned procedure, $\Delta\chi_{DIA}$ may be obtained for oxides with high crystal symmetry as described before.

Recently, micro-gravity experiments were conducted in order to improve sensitivity of $\Delta\chi$. Here the fiber was completely excluded from the system [13]. Experiments were recently performed at MGLAB [Micro gravity laboratory of Japan]; here a highly-qualified μG condition of 2.0×10^{-3} Gal was produced for 4.5 sec. According to eq.(2), it is essential to increase duration of μG in order to increase sensitivity of $\Delta\chi$[8]. When rotational oscillation is realized in orbital laboratory, sensitivity of $\Delta\chi$ measurement will be increased almost limitlessly. Hence, $\Delta\chi_{DIA}$ value may be obtained for most of the highly-symmetrical crystals in μG condition. However it is also important to develop high-sensitive $\Delta\chi$ measurement in normal-gravity condition, because the method is far easier to perform compared to the measurements in μG. It is noted that compiling of $\Delta\chi_{DIA}$ data at a level of 10^{-10} emu/g may proceed rapidly by adopting the present method.

An upper limit of available field intensity has increased considerably during the last several decades [14]. Novel applications on diamagnetic alignment may be realized when small $\Delta\chi$ values of inorganic materials with highly symmetric structures are obtained.

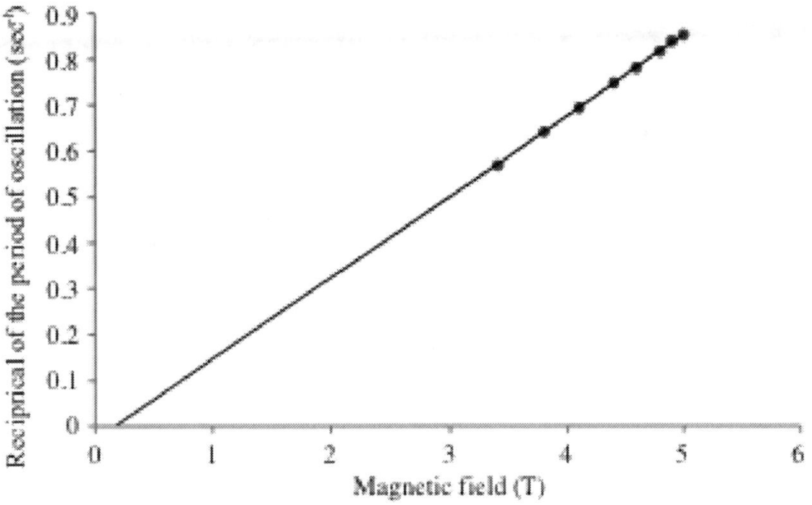

Figure 2. τ^{-1}-B relationship of standard sample.

4. Conclusion

An experiment to determine $\Delta\chi$ values using magnetic field of 5.0 T was realized by setting a conventional apparatus to observe harmonic-oscillation of a sample in a cryogen-free magnet. High sensitivity of $\Delta\chi$ measurement was achieved; $\Delta\chi$ value of 5×10^{-10} emu/g was accurately detected. A setup for low temperature experiment is now being developed in order to separate the contribution of $\Delta\chi_{DIA}$ from $\Delta\chi_{PARA}$. The presented system is effective for investigating $\Delta\chi_{DIA}$ values of symmetric crystals, which are expected to be smaller than 1×10^{-9} emu/g.

References

[1] Various papers appering in *Proc. Int. Symp. New Magneto-Sci. Tsukuba* 1999 Jpn. Sci. and Tech. Corp., NIMS
[2] Sazaki G, Yoshida E, Komatsu H, Nakada T, Miyashita K and Watanabe K 1997 Effects of a magnetic field on the nucleation and growth of protein crystals *J. Cryst. Growth* **173** 231-234
[3] Maret G and Dransfeld K 1985 Biomolecules and polymers in high steady magnetic fields *Top. Appl. Phys.* **57** 143-204
[4] Uyeda C 1993 Rotational Motion of Inorganic Crystals Induced in the Magnetic Field *Jpn. J. Appl. Phys.* **32** L268-L270
[5] Uyeda C 1993 Diamagnetic anisotropies of oxide minerals *Phys. Chem. Miner.* **20** 77-81
[6] Uyeda C, Tanaka K and Takashima R 2004 Stabilization of the Position of a Solid Body Isolated in Micro-Gravity and Field-Induced Rotational-Oscillation Caused by Diamagnetic Anisotropy *Jpn. J. Appl. Phys.* **42** L1226-L1228
[7] Uyeda C, Takashima R and Tanaka K 2005 Magneto-rotation of nonmagnetic micro-crystals caused by diamagnetic anisotropy *Appl. Phys. Lett.* **86** 094103
[8] Uyeda C and Abe T 2006 Alignment of nonmagnetic solid grains that essentially require high magnetic field; alignment of hexagonal ice I_h *J. Phys., Conf. Ser.* **51** 466-469
[9] Uyeda C, Abe T and Takashima R 2006 Magnetic Anisotropy of Ordinary Crystals without Spontaneous Magnetic Moments Detected by Rotational Oscillations Induced by a Field of a Horseshoe Magnet *Jpn. J. Appl. Phys.* **45** L268-L271
[10] Schmidt V, Hirt AM, Rosselli P and Martín-Hernández F 2007 Separation of diamagnetic and paramagnetic anisotropy by high-field, low-temperature torque measurements *Geophys. J.*

Int. **168** 40-47

[11] Guputa R 1983 Diamagnetism (Springer-Verlag, Berlin, 1983) *Londert Börnstein, New Serious, Vol. II/* 16

[12] Krishnan KS and Banerjee S 1936 Investigations on Magne-Crystallic Action. IV. Magnetic Behaviour of Paramagnetic Ions in the S-State in Crystals *Philos. Trans. R. Soa. London. Ser.* A. **235** 343-366

[13] Uyeda C, Mamiya M, Takashima R, Abe T, Nagai H and Okutani T 2006 Diamagnetic Anisotropy Detected by a Magnetic Oscillation in Drop Tube without Suspending Crystals with Fiber *Jpn. J. Appl. Phys.* **45** L124-L127

[14] for example, Kimura S, Takeuchi T, Okunishi K, Hagiwara M, He Z, Kindo K, Taniyama T and Itoh M 2008 Novel Ordering of an S=1/2 Quasi-1d Ising-Like Antiferromagnet in Magnetic Field *Phys. Rev. Lett.* **100** 057202

3rd International Workshop on Materials Analysis and Processing in Magnetic Fields (MAP3) IOP Publishing
Journal of Physics: Conference Series **156** (2009) 012024 doi:10.1088/1742-6596/156/1/012024

Attempt to Improve Sensitivity in Measuring Diamagnetic Anisotropy by Increasing Duration of Rotational Oscillation in Microgravity

C Uyeda[1], S Kano[1], K Hisayoshi[1], T Kashiwagi[2], T Kida[2] and M Hagiwara[2]

[1]Institute of Earth and Space Science, Graduate School of Science, Osaka University, Osaka, Japan

[2]KYOKUGEN (Center for Quantum Science and Technology under Extreme Conditions), Osaka University, Osaka, Japan

E-mail: uyeda@ess.sci.osaka-u.ac.jp

Abstract. Technical basis to detect small diamagnetic anisotropy $\Delta\chi_{DIA}$ is established in a method using microgravity μG. Sensitivity of $\Delta\chi$ can be improved by increasing length of measurable period of rotational oscillation of magnetically stable axis with respect to field direction. In order to achieve the above condition, position of the diamagnetic sample should be stabilized in static field area, which was realized by introducing a magnetic circuit with high field homogeneity. Function of a mobile sample stage was effective in minimizing translational motion of sample. Accordingly, magnitude of measurable anisotropy may be decreased to a level of 10^{-10} emu/g. $\Delta\chi_{DIA}$ of materials with high crystal symmetry, namely hexagonal ice, ZnO, SiC and $BaTiO_3$, can be obtained by the above improvement. Magnetically active property due to magnetic anisotropy may be recognized for almost all the solid materials when sensitivity reaches the above level.

1. Introduction

Property of conventional magnet has been commonly recognized by rotational motion of a compass caused by terrestrial field. Similar oscillation was reported recently on a number of ordinary solids free of spontaneous moment at a low field of a hand magnet [1]; the oscillation was observed for popular materials such as biotite, calcite, forsterite, graphite, gypsum or orthoclase. The results indicate that many of the ordinary solid are magnetically active at practical low field without a use of high-costing superconducting magnet, in contrary to a common recognition. The rotation is caused by anisotropy of magnetic susceptibility $\Delta\chi$, which is generally expressed as sum of diamagnetic anisotropy $\Delta\chi_{DIA}$ and paramagnetic anisotropy $\Delta\chi_{PARA}$ [2]; $\Delta\chi_{DIA}$ originates form anisotropy of localized electron distribution, whereas $\Delta\chi_{PARA}$ is derived form magnetic impurity ion contained in crystal. It is essential to study the origin of $\Delta\chi$ at atomic level in order to design an industrial process in producing a functional material; from an practical point of view, it is desirable that such process is achieved at a low field intensity[3,4].

A model was proposed to explain the origin of published $\Delta\chi_{DIA}$ recently compiled for oxide crystals with various structures[5]. In the model, an unixial anisotropy $\Delta\chi_{BD}$ was assigned to individual bonding orbital with the bond direction being assumed as a magnetically unstable axis. This assignment was compatible with spatial distribution of orbital. The above model was consistent with published values of various oxides provided that $\Delta\chi_{DIA}$ was larger than $1x10^{-9}$emu/g. According to numerical fitting between model and published $\Delta\chi_{DIA}$ values, $\Delta\chi_{BD}$ was assigned as $-1.1x10^{-30}$emu for a hydrogen bond, $-3.7x\ 10^{-30}$ for a T-O bond of tetrahedral [TO_4] unit and $-0.32x10^{-30}$emu for M-O bond of octahedral [MO_6] unit. It was expected that $\Delta\chi_{DIA}$ of unmeasured crystal could be predicted by using the above $\Delta\chi_{BD}$ values. Oxides with high crystal symmetry, such as a wurtzite-, a rutile- or a perovskite structure, were expected to posses small $\Delta\chi_{DIA}$ that ranges between 10^{-10}-10^{-11}emu/g.

$\Delta\chi_{DIA}$ of inorganic oxides has been considered to be negligibly small to be detected by a conventional method [6,7]. Small $\Delta\chi_{DIA}$ of many popular oxides were detected by a method using harmonic-oscillation of a sample induced by static horizontal field below B=2T. This oscillation was achieved in a condition where restoration torque of a fiber suspending the sample was much small compared to torque of magnetic anisotropy. $\Delta\chi_{DIA}$ of the level of 1×10^{-9} emu/g was obtained by the above method [8]. However it was impossible to detect $\Delta\chi_{DIA}$ of the above-mentioned oxides with high crystal symmetry by this system.

It was recently proposed that sensitivity of $\Delta\chi$ may be improved limitlessly by increasing period of the above-mentioned oscillation τ in microgravity μG, where restoration torque of a fiber suspending the crystal was excluded from measuring system. A preliminary experiment was performed in microgravity μG in order to examine the above principle [9]. However, sensitivity remained at level of 10^{-9}emu/g that was comparable to sensitivity obtained from rotational oscillation performed in terrestrial gravity. This was because a sample was ejected from field centre by field-gradient force; measurable length of τ remained below 1.3s. In the present work, a neodymium-iron-boron magnet that produced homogenous field was introduced to stabilize sample position. Possibility of improving $\Delta\chi$ sensitivity is discussed based on increased duration of rotational oscillation at field centre.

2. Experimental

Microgravity was produced in a drop capsule at Micro-gravity Laboratory of Japan MGLAB (Toki, Gifu, Japan). Duration of microgravity T_d was 4.5s in a single drop. Instability of μG was less than $5x10^{-4}$G;[13] no motions due to g-jitter were observed in the experiment. An apparatus developed to measure magnetic anisotropy in μG condition [9] is shown in Fig.1. A permanent magnetic circuit (NEOMAX X-1466) is used to generate a homogeneous field of that produced magnetic field of B = (1.15700 ± 0.00002) Tesla in a region of $5x10^{-2}$ m in diameter. A single crystal of α-quartz was placed on a mobile sample stage, which was initially at the center of the area of the homogeneous field.

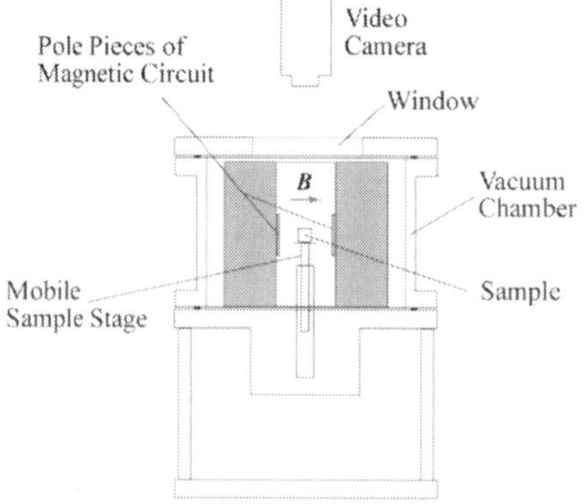

Figure1. Sectional view of apparatus developed to observe motion of sample caused by magnetic anisotropy [9].

3. Results and Discussions

Oscillation of magnetically stable axis with respect to ***B*** was observed for α-quartz as shown in Fig.2. τ value obtained from the images was consistent with the values calculated by the following equation that was deduced from a rotational equation.

$$\tau' = 2\pi\sqrt{\frac{I}{N\Delta\chi}}B^{-1}\left\{1+\frac{1}{4}\sin^2\frac{\theta_0}{2}+...\right\}. \quad (1)$$

I and *N* denote moment of inertia and mass of sample, respectively. θ_0 is amplitude angle. Magnetic anisotropy of α-quartz sample was obtained as $\Delta\chi_{exp}$ = (2.2x0.3)x10^{-9} emu/g, by inserting observed values of τ =1.92±0.1s, I/m=(0.0417±0.006) cm^2, θ_0 =(47±2)° and above-mentioned *B* value in eq.(1). The value is consistent with a published value $\Delta\chi_{pub}$ = 2.0x10^{-9} emu/g. Uncertainty of $\Delta\chi_{exp}$ mainly derives from uncertainty of τ and *I/m*. Here oscillation was measured for about 1.3 cycles, and uncertainty of τ was evaluated from time resolution of a video camera in the present report. It is noted that oscillation was observed as long as microgravity continued. This was possible because the high homogeneity of magnetic circuit stabilized position of the sample. Function of the above-mentioned sample stage minimized initial velocity of sample; hence sample remained inside region of homogeneous field.

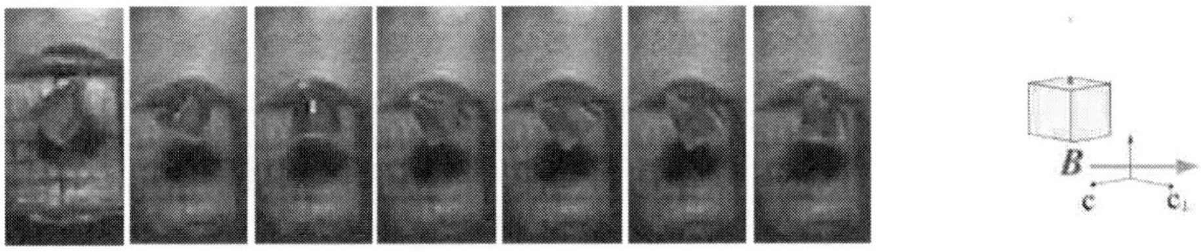

Figure 2. Visual images of α-quartz single-crystal taken from vertical directions as described in Figure 1. The images are arranged in time sequence from left to right. Image at the left end was taken just after microgravity achievement of μG. Time interval between images was 0.667sec. Magnetic field was applied in vertical direction in all the images. Two magnetically stable axes of a sample were initially inclined 47° with respect to filed as shown in right portion of figure.

It is deduced from eq.(1) that smaller Δχ can be obtained by increasing *B* or *τ*, or else by decreasing *I*. It is difficult to increase *B* or decrease *I* in the present setup. Whereas length of τ can be increased to T_d = 4.5 sec at MGLAB if the sample position is stabilized as shown in Fig.2. This enhancement of τ realizes measurement at the level of 10^{-10}emu/g as shown in Fig.3. The sensitivity may be improved to a level of 10^{-11}emu/g in a parabolic flight where measurable τ is enhanced to 20 seconds.

Oscillation in microgravity reproduces a primitive motion deduced from a primitive equation of magnetic torque. Rotational motion of spontaneous moment has induced inventions of important devices in the past, irrespective of the refined theories in modern magnetism. In a same sense, motions of ordinary crystals, as shown in Fig.2, may induce new applications in various research fields. Compiled magnetic anisotropy data show that most of the micron-sized crystals dispersed in fluid show almost full alignment below 10 Tesla [3]. It is noted that origin of anisotropy that induce alignment is not completely solved as yet, which should be a theme of quantum chemistry or solid state physics.

The above oscillation was expected to be preserved at T=1000K for most of the oxides [10]. It is a characteristic proposed for diamagnetic and paramagnetic materials in general, which is not widely recognized as yet; spontaneous moments are usually diminished above 1000K. Application of magnetic alignment in natural conditions as well as in industrial application may increase by adopting this property. For example, alignment of pre-solar grains may occur in envelops of red giant stars (T =1000~2000K). Direction of stellar field, which is a major factor that control star evolutions, is determined from polarization caused by the grain alignment. In this sense, the above-mentioned $\Delta\chi$ measurements should be performed in the high temperature region, in order to confirm that efficiency of alignment is preserved up to melting temperature of a solid material.

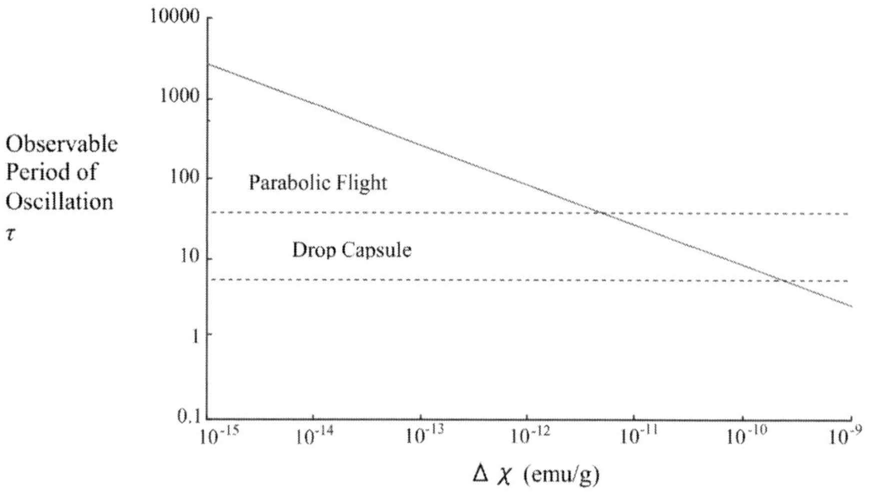

Figure 3. Relationship between observable τ and magnetic anisotropy $\Delta\chi$ calculated for cubic sample released in micro-gravity which has edge length of 0.3 cm. Relationship is described for field intensity of 1.0 Tesla.

4. Conclusion

Magnetic anisotropy can be measured over a wide range between 10^{-5} emu/g and 10^{-11} emu/g in terms of a single setup based on a rotational oscillation of sample in microgravity. Accordingly, efficiency of magnetic alignment may be quantitatively examined for most of the oxide crystal. Almost full alignment is expected at field intensity below $B = 10$T for micron-sized crystals dispersed in fluid at room temperature.

References

[1] Uyeda C, Abe T and Takashima R 2006 *Jpn. J. Appl. Phys.* **45** L268
[2] Uyeda C, Tanaka K and Takashima R 2004 *Jpn. J. Appl. Phys.* **42** 980
[3] Uyeda C, Takashima R and Tanaka K 2005 *Appl. Phys. Lett.* **86** 094103
[4] Various papers apperaing in Proc. Int. Symp. New Magneto-Sci. Tsukuba (Jpn. Sci. and Tech. Corp., NIMS, 1999)
[5] Uyeda C 1993 *Phys. Chem. Miner.* **20** 77
[6] Guputa R 1983 *Diamagnetism* (Springer-Verlag, Berlin) Londert Bornstein, New Serious, Vol. II/ 16
[7] Maret G and Dransfeld K 1985 *Top. Appl. Phys.* **57** 143
[8] Uyeda C 1993 *Jpn. J. Appl. Phys.* **32** L268
[9] Uyeda C, Mamiya M, Takashima R, Abe T, Nagai H and Okutani T 2006 *Jpn. J. Appl. Phys.* **45** L124
[10] Uyeda C, Abe T and Takashima R 2006 *Jpn. J. Appl. Phys.* **45** L268

Optical Extinction Change of Magnetic Nanoparticle Suspension under Pulsed Magnetic Field

M Suwa and H Watarai

Department of Chemistry, Graduate School of Science, Osaka University

E-mail: msuwa@chem.sci.osaka-u.ac.jp

Abstract. We investigated the optical extinction change of γ-Fe_2O_3 magnetic nanoparticle suspension induced by a pulsed magnetic field. The Extinction decrease accompanied by a pulsed magnetic field showed clear relationship with the magnetization of magnetic nanoparticle. However, there observed a different response to the magnetic field between increasing and decreasing of the magnetic field. This indicated that the extinction change followed the time variation of magnetic field with some time constant. The time constant was estimated to be 4.5 μs, but it was different from either of rotational Brownian relaxation (0.23 μs) and Neel relaxation (10 ns) of individual particle. It was thought that the optical density decrease might be attributed to the decrease of Rayleigh scattering cross section by the rotation of the magnetic nanoparticle aggregates....

1. Introduction

Magnetic nanoparticle (MNP) suspension, which is known as ferrofluid, has attracted wide interest in scientific and engineering fields. The MNP has been used for various purposes, such as a lubricant and seals of rotating shaft [1], a contrast medium of MRI [2], a destroyer of cancer tumors by magnetic hyperthermia [3], a substrate for protein purification [4]. In colloid science, MNP has been utilized as a model magnetic colloidal system, because they have permanent magnetic moment. Attractive or repulsive interaction between MNPs determined the stability and the aggregation [5]. The study of dipolar interactions on the aggregation of MNP will provide a fundamental insight in the behavior of colloidal particles. The aggregation of MNPs under magnetic field has been studied by measuring photo-absorption spectra [6], complex magnetic susceptibility [7], electron microscope [8] and magneto-optical effect [9].

In the present study, we observed the change of optical extinction of MNP suspension accompanied with pulsed magnetic field. The time course of the optical density change was analyzed from the view point of magnetization. The comparison with Langevin function, which describes the magnetization behavior of MNP suspension, suggested that this phenomenon could be attributed to the aggregation of MNPs.

2. Materials and Method

2.1. Ionic maghemite nanoparticle aqueous dispersion

γ-Fe_2O_3 nanoparticle (MNP) was synthesized by the method described by van Ewijk et al.[10]. All regents were used as received without further purification. Two milliliter portion of 25 % NH_3 was added into 50 mL of 0.04 M $FeCl_2$ and 0.02 M $FeCl_3$ aqueous solution, and then, black precipitate was formed. After the mixture was vigorously agitated, the precipitate was gathered by using a permanent magnet and the supernatant was removed. Four milliliter of 2 M HNO_3 aqueous solution and 6 mL of 0.35 M $Fe(NO_3)_3$ was added to the black precipitate and the mixture was boiled for 10 min. During boiling, the color of the precipitate changed to reddish yellow. After washing several times with 2 M HNO_3, the precipitate was dispersed in 5 mL of water purified by milli-Q system. MNP dispersion was diluted to $3.3 \times 10^{-16} - 1.2 \times 10^{-14}$ particles / L before measurement. The concentration of MNP was estimated from its magnetic susceptibility measured by magnetic balance (MSB-AUTO, Sharewood Scientific) under the magnetic field of 0.45 T.

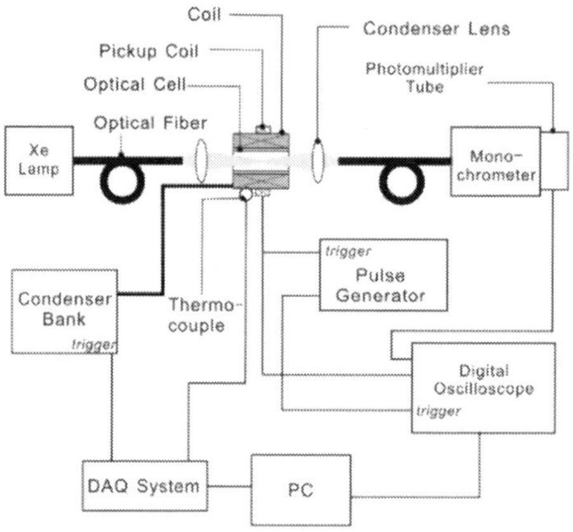

Figure 1. The experimental setup for observation of optical density change accompanied with pulsed magnetic field.

2.2. Photo-extinction measurement under pulsed magnetic field

Figure 1 illustrates the experimental setup to observe the optical transmission change under a pulsed magnetic field. The pulsed magnetic field was generated by a 2000 μF capacitor bank (MAG-2520L-3A, Toyojiki Co., ltd.) and a 16.4 μH hand made solenoidal coil, whose bore size was 8 mmϕ diameter × 16 mm length. The applied voltage on the capacitor was between 50 V and 500 V. The incident and transmitted lights were guided to a cell in bore of coil and a monochrometer (MD200, Unisoku), respectively, by optical fibers. A cylindrical glass cell, whose optical length was 20 mm, was set in the bore of the coil. The signal from photomultiplier (PM) tube (R2429, Hamamatsu Photonics) was stored on a digital oscilloscope (LT-262, Lecroy) synchronized with a TTL signal from a pulse generator (PG-230, IWATSU) triggered by a signal from a pickup coil. The photo-current signal from the PM was converted to voltage and then the voltage in the absence of magnetic field, V_0, was set to be 1.0 V. The oscillogram was transferred to PC and averaged for 10 ~ 50 measurements. The optical density change, ΔA, was obtained from the photo-voltage change, ΔV, by using the following equation:

$$\Delta A(t) = -\log\left(1 + \frac{\Delta V(t)}{V_0}\right) \quad (1)$$

The temperature of the sorenoidal coil was monitored with a thermocouple and, in order to avoid breaking of the coil, magnetic pulse was controlled not to increase the temperature above 26 °C by a data acquisition system (NI-compact DAQ, National Instruments) programmed with LabVIEW® software.

Journal of Physics: Conference Series **156** (2009) 012025 doi:10.1088/1742-6596/156/1/012025

2.3. Magnetic field measurement
Generated magnetic field was evaluated from Faraday effect [11]. Faraday rotation angle of the same glass cell filled by water was measured. Laser light of 640 nm in wavelength after passing a polarizer was irradiated into the cell put in the bore of the coil. Faraday rotation angle was measured from the change of light intensity through an analyzer accompanied with pulsed magnetic field.

3. Results and Discussion
Figure 2 shows the transmission electron microphotograph of the MNP synthesized in the present study. The average diameter was 9.1 ± 3.0 nm. The MNP dispersed in distilled water was very stable without further treatment. Figure 3(A) indicates the optical density decrease at 390 nm accompanied with pulsed magnetic field. The optical density was steeply decreased at the start of applying the field, instantly attained to the constant value, and then gradually increased with the decrease of the magnetic field. The magnetic field was plotted against time in Fig. 3(B). It was found that the time course of the extinction change was different from that of magnetic field. Magnetic field strength increased with the voltage on the condenser bank, while the transmitted light intensity change was not proportional to the magnetic field change.

The wavelength dispersion of the maximum value of optical density change was measured as shown in Fig. 4. The conventional optical extinction spectrum measured with a spectrophotometer indicates the typical Rayleigh scattering pattern, which shows reciprocal dependence on wavelength. The optical density change showed a good correlation with the optical extinction spectrum in the absence of pulsed magnetic field. Therefore, the extinction change induced by magnetic field could be contributed to the variation of the scattering efficiency due to the rotational motion or aggregation of MNP under pulsed magnetic field.

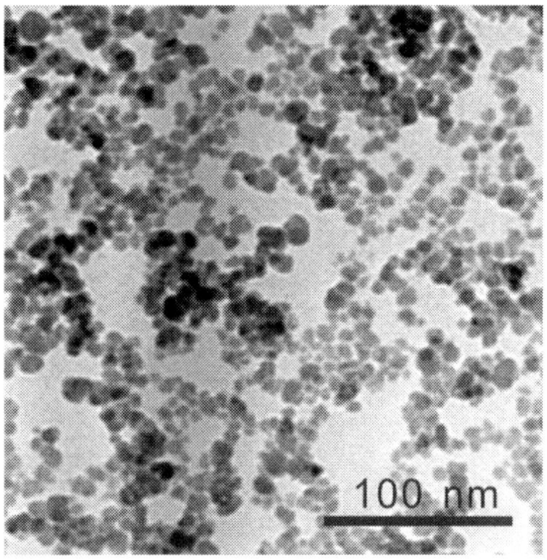

Figure 2. Transmission electron microphotograph of MNPs

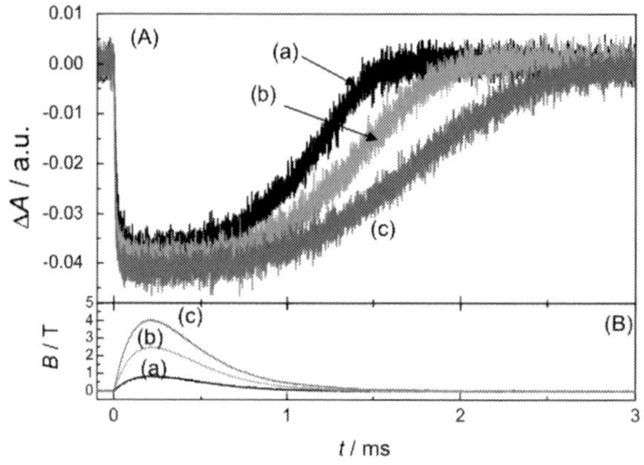

Figure 3. Typical time courses of optical density (A) and magnetic field measured from Faraday rotation angle (B): The voltages charged on the condenser bank were 100 V (a), 300 V (b) and 500 V (c). The observed wavelength was 390 nm, each oscillogram was the average of 50 measurements. The concentration of MNP was 1.6×10^{15} particles L^{-3}.

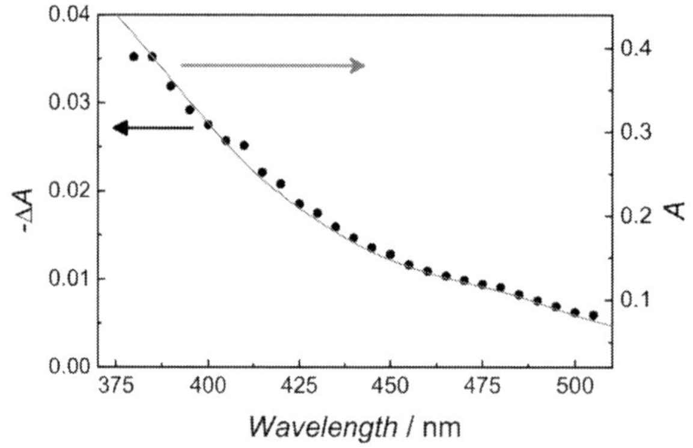

Figure 4. The spectra of optical extinction under no magnetic field (line) and that of optical density change pulsed magnetic field (filled dots). The voltage on the condenser bank was fixed to 100 V.

For elucidating the observed extinction change, it should be useful to consider the superparamagnetic magnetization behavior of MNP, because that is very sensitive to magnetic field and strongly related to the motion of MNP. In magnetic field, the magnetic moment of MNP, m_{p}, tends to orient along the field direction and the magnetization can be described by Langevin function:

$$\frac{M}{M_{\infty}} = \mathrm{L}(B) = \coth(\beta B) - \frac{1}{\beta B}$$
$$\beta = \frac{m_{\mathrm{p}}}{k_{\mathrm{B}} T} \tag{2}$$

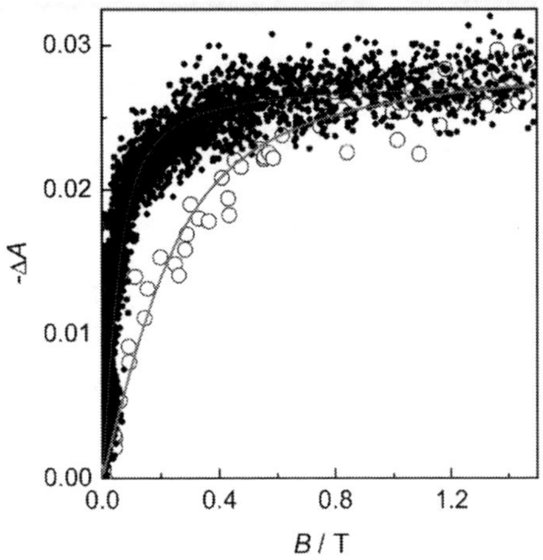

Figure 5. Optical density change *vs.* magnetic field: The voltage on the condenser bank was 500 V. The solid curve was fitted by Eq. (4).

where M_∞ is the saturation magnetization, m_p the magnetic moment of MNP, T the temperature, and k_B the Boltzmann's constant. The magnetic moment of MNP can be represented as follows:

$$m_p = M_s V \qquad (3)$$

where M_s and V is the spontaneous magnetization and the volume of MNP. The magnetic momentum of examined γ-Fe_2O_3 nanoparticle could be estimated to 1.4×10^{-19} Am^2 by substitution of M_s of 3.5×10^5 Am^{-1} [12] and V of 3.9×10^{-25} m^3 into eq.(3). So as to present clearly the relationship between magnetization and the optical density change, ΔA was plotted against the magnetic flux density in Fig. 5. The open circles indicate the optical density change in the increasing time region of pulsed magnetic field, and the filled dots show that for the decreasing time region of the magnetic field. A constant sampling frequency of the oscilloscope is responsible for the difference of the number of data between two processes. As shown in Fig. 3(B), the magnetic field quickly increased and comparatively slowly decreased, therefore, the sampling number during increase of magnetic field was smaller than that during decrease. Each plot was fitted by the following equation:

$$\Delta A = \Delta A_s \, \mathrm{L}(B) \qquad (4)$$

where ΔA_S was the saturated value of ΔA, as indicated by solid lines. The fitted curves seem to be in good agreement with the experimental data. However, it was found that there was a difference between increasing and decreasing time regions. Therefore, the intensity change was attributed to some relaxation behavior that could not follow the pulsed magnetic field.

The relaxation behavior of the MNP magnetization is attributed to two kinds of mechanisms.[5] One is the Brownian relaxation. The magnetic moment of MNP particles suspended in water can orient along the magnetic field accompanied by the rotation of the particle. Therefore, rotational Brownian motion is responsible for the relaxation behavior. The Brownian relaxation time, τ_B, can be written by the following equation:

$$\tau_B = \frac{3\eta V_H}{k_B T} \qquad (5)$$

where η is the viscosity of medium and V_H is the hydrodynamic volume of particle. Another one is Néel relaxation. The anisotropic magnetic moment of MNP may also affect its internal orientation

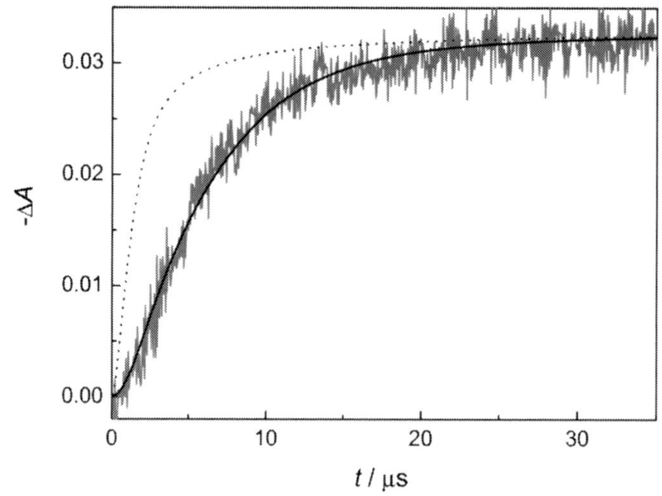

Figure 6. The time course of ΔA at the start of pulsed magnetic field in the case that the voltage charged on the condenser bank was 500 V: The dashed line and solid line indicate the simulations calculated by using eq. (8) for τ= 10 ns and 4.5 μs, respectively.

overcoming the energy barrier constituted by the anisotropy energy. The relaxation time for this process can be expressed as:

$$\tau_N = \tau_0 \exp\left(\frac{KV_P}{k_B T}\right) \tag{6}$$

where K is the anisotropy constant, with τ_0 usually having a value of 1 ns. [13] In the present system, τ_B and τ_N were estimated to be 0.23 μs and 10 ns, respectively, by substituting 2.5×10^4 $\mathrm{Jm^{-3}}$ of the anisotropy constant [14]. Since both processes are present simultaneously, an effective relaxation time of magnetization, τ_{eff}, will be shorter than both relaxation times and can be expressed as:

$$\tau_{eff} = \frac{\tau_B \tau_N}{\tau_B + \tau_N} \tag{7}$$

Substitution of the values of τ_B and τ_N into eq. (7) yields about 10 ns as the value of τ_{eff} and therefore it could be said that Néel relaxation controlled the rate of magnetization.

Figure 6 shows the expansion of the time course of the optical density change. In this time region, the magnetic field could be approximated to be proportional to time. We simulated the optical density change, assuming that ΔA had linear dependence on the magnetization of MNP, that is Langevin function, and that the magnetization could not follow the variation of the pulsed magnetic field. This assumption allowed us to calculate the time course of ΔA by a convolution integral as represented by following expression:

$$\Delta A(t) = \int_0^t \mathrm{L}(\alpha t_s)\Delta A_s \exp\left(-\frac{t-t_s}{\tau}\right)\frac{\mathrm{d}t_s}{\tau} \tag{8}$$

where α was the proportional coefficient of the magnetic field to time. The dashed line in Fig. 6, which indicates the calculated result in the case of τ = 10 ns, did not agree with the experimental results. Therefore, we determined the relaxation constant so as to minimize the squared value of the difference between the experimental results and the calculated results. Thus, the bold line in Fig. 6 and the value of 4.5 μs were obtained. The calculated results were in good agreement with the experimental results. Therefore, it was suggested that the rate controlling step of the extinction change is not the magnetization step but any other phenomenon after the magnetization should be considered.

One of the most likely explanations for the optical extinction change is that the decrease of the scattering cross section due to the orientation of individual MNP which has anisotropic shape, that is spheroid or ellipsoid. However, from the TEM observation, MNPs were rugged but seem not to be anisotropic shape, and the Brownian relaxation, τ_B, for the present sample was estimated to be much smaller than the observed value from the time course of the extinction change. Furthermore, τ_N was much smaller than τ_B, because the permanent magnetic moment of MNP could orient within the particle without any rotation of the particle itself. So it is difficult to attribute the extinction change to the orientation of individual MNPs. Finally, the aggregate of MNPs should be considered. MNPs could be spontaneously aggregated by magnetic dipolar-dipolar interaction [5, 8]. In the absence of external magnetic field, MNP aggregates were randomly oriented. Once pulsed magnetic field was exerted, the magnetic moment oriented to the magnetic field by Néel process and then the aggregates rotated to minimize the potential energy. Potential energy should be minimum in the case that the longest axis of aggregate orients along to the magnetic field, and then the scattering cross section decreases.[15] In this case, the relaxation time depends on the rotational Brownian motion and the time constant can be described by eq. (5). The hydrodynamic diameter could be estimated to be 25 nm from the obtained time constant (4.5 μs). This value seems reasonable assuming that the MNPs formed dimmer or trimmer. The other considerable reason is reversible aggregation and dissociation. The magnetic interaction between the MNPs should be changed by the application of magnetic field due to the rotation of magnetic moments. In the case that MNP dimmer aligned perpendicular to magnetic field, the interaction of magnetic moments will be repulsive after Néel relaxation. On the other hand, in the case that the MNPs aligned along the magnetic field, the interaction will be attractive. In the former case, scattering efficiency will become small because that is proportional to the square of volume of particle following Rayleigh scattering mechanism. In the latter case, the MNPs will form chain like structure along the magnetic field, and then the scattering cross section will decrease. In both case, the process needs the translational motion of MNP. However, the extinction change had linear relationship with Langevin function as demonstrated by Fig. 6, and the translational motion attributed to magnetic dipole-dipole moment interaction should be more complicated. Therefore, it is difficult to ascribe the extinction change to the dissociation and aggregation dynamics. As aforementioned discussion, it was suggested that the extinction change could be contributed to the decrease of the scattering cross section by orientation of MNP aggregation. In order to clarify the origin of the extinction change, the dependence of the relaxation time constant on the concentration of MNP, on the ionic strength, on the chemical property of surface, and on the particle size should be investigated.

4. Conclusion

We found the optical extinction change of MNP suspension under pulsed magnetic field. The extinction change had the strong relationship with the magnetization of MNP, because the optical density change could be fitted by Langevin function. However, a hysteresis was observed in the extinction change. The assumption that the optical density change has a linear relationship with Langevin function and it takes τ to relax the change allowed us to use the convolution integral and simulate the optical density change. The simulation was in good agreement with the experimental result. From this evidence, it was suggested that the aggregate of MNP is ascribed to the extinction change.

5. Acknowledgement
The authors would like to thank Dr. Satoshi Fujii, Graduate School of Science, Osaka University for his help with TEM measurement. This work was supported by Grant-in-Aid for Young Scientists (B) (18750063) of Japan Society for the Promotion of Science.

Reference
[1] Ellis A B, Geselbracht M J, Johnson B J, Lisensky G C, Robinson W R 1993 *Teaching General Chemistry, A Material Science Comparison*, (Washington DC: American Chemical Society)
[2] David E. Sosnovik, Matthias Nahrendorf, Ralph Weissleder 2008 Magnetic nanoparticles for MR imaging: agents, techniques and cardiovascular applications *Basic Res. Cardiol.* **103**, 122-130
[3] Pankhurst Q A, Connoly J, Jones S K, Dobson J 2003 Applications of magnetic nanoparticles in biomedicine *J. Phys. D: Appl. Phys.* **36** R167-R181
[4] Berry C, Curtis A 2003 Functionalisation of magnetic nanoparticles for applications in biomedicine *J. Phys. D: Appl. Phys.* **36** R198
[5] Eberbeck D, Wiekhorst F, Steinhoff U, Trahms L 2006 Aggregation behaviour of magnetic nanoparticle suspensions investigated by magnetorelaxometry *J. Phys. Condens. Matter.* **18** S2829-S2864
[6] Petersson K, Ilver D, Johansson C, Krozer A 2006 Brownian motion of aggregating nanoparticles studied by photon correlation spectroscopy and measurements of dynamic magnetic properties *Anal. Chim. Acta* **573-574** 138-146
[7] Kolkkenburg M, Erné B H 2006 Comparison of reversible and irreversible dipolar assemblies in a ferrofluid *J. Magn. Magn. Mater.* **306** 85-91
[8] Butter K, Bomans P H H, Frederik P M, Vroege G J, Philipse A P 2003 Direct observation of dipolar chains in iron ferrofluids by cryogenic electron microscopy *Nature Mater.* **2** 88-91
[9] Aurich K, Nagel S, Glöckl G, Weitschies W 2007 Determination of the Magneto-Optical Relaxation of Magnetic Nanoparticles as a Homogeneous Immunoassay *Anal. Chem.* **79** 580-586
[10] van Ewijk G A, Vroege G J, Philipse A P 1999 Convenient preparation methods for magnetic colloids *J. Magn. Magn. Mater.* **201** 31-33
[11] Barron L 2004 Molecular Light Scattering and Optical Activity Second Edition (Cambridge: Cambridge University Press)
[12] Wilhelm C, Gazeau F, Roger J, Pons J N, Salis M F, Perzynski R, Bacri J C 2002 Binding of biological effectors on magnetic nanoparticles measured by a magnetically induced transient birefringence experiment *Phys. Rev. E* **65** 031404
[13] Hrianca I, Caizer C, Schlett Z 2002 Dynamic magnetic behavior of Fe_3O_4 colloidal nanoparticles *J. Appl. Phys.* **92** 2125-2132
[14] Domínguez M, Ortega D, Garitaonandía J S, Litrán R, Barrera-Solano C, Blanco E, Ramírez-del-Solar M 2008 Magneto-optic Faraday effect in maghemite nanoparticles/silica matrix nanocomposites prepared by the Sol–Gel method *J. Magn. Magn. Mater.* **320** e725-e729
[15] H. C. van de Hulst 1957 *Light Scattering by Small Particles* (New York: Dover Publications Inc.)

Journal of Physics: Conference Series **156** (2009) 012026 doi:10.1088/1742-6596/156/1/012026

Effects of Magnetic Processing on Electrochemical and Photoelectrochemical Properties of Electrodes Modified with Nanoclusters of a Phenothiazine-C_{60} Linked Compound

H Yonemura[1]*, Y Wakita[2], S Moribe[2], S Yamada[1,3], Y Fujiwara[4] and Y Tanimoto[4, 5]

[1] Department of Applied Chemistry, Faculty of Engineering, Kyushu University, Fukuoka 819-0395, Japan

[2] Department of Materials Physics and Chemistry, Graduate School of Engineering, Kyushu University, Fukuoka 819-0395, Japan

[3] Center for Future Creation, Kyushu University, Fukuoka 819-0395, Japan

[4] Department of Mathematical and Life Sciences, Graduate School of Science, Hiroshima University, Higashi-Hiroshima, Hiroshima 739-8526, Japan

[5] Department of Mathematical and Life Sciences, Graduate School of Science, Hiroshima University, Higashi-Hiroshima, Hiroshima 739-8526, Japan

E-mail: yonemura@mail.cstm.kyushu-u.ac.jp

Abstract. Effects of magnetic processing on morphological, electrochemical, and photoelectrochemical properties of electrodes modified with nanoclusters of a phenothiazine-C_{60} linked compound with four methylene group (**Ph(4)C_{60}**) were examined in the absence and presence of magnetic processing with three different magnetic environments due to strong magnetic field. The AFM measurements indicated that the morphologies of nanostructures of **Ph(4)C_{60}** varied with magnetic enviroments as comparison with that in the absence of magnetic processing. At top position (5.6 T ; -940 T^2/m) with hypogravity, large spherical nanoclusters (60~70 nm diameter) were observed as comparion with those (ca. 20 nm diameter) in the absence of magnetic processing. At middle positon (15 T ; 0 T^2/m) with normal gravity, the fiber-like nanostructure was observed. At bottom position (9.8 T ; +1070T^2/m) with hypergravity, the rod-like nanostrucure was observed. The interesting results might be ascribed to the different solvent properties due to the different rates of evaporation of two solvents in the toluene-acetonitrile mixed solvent during drying process under various magnetic environments. First reduction peaks due to C_{60} moiety of **Ph(4)C_{60}** nanostrucures in the presence of magnetic processing at three different positions were negative-shifted as comparison with that in the absence of magnetic processing. Potential dependencies of the photocurrents of the electrodes modified with **Ph(4)C_{60}** nanostrucures in the presence of magnetic processing at three positions were also different from that in the absence of magnetic processing. The magnetic field effects in AFM, and electrochemical and photoelectrochemical measurements are most likely ascribed to the difference of the reduction potentials due to C_{60} clusters between the absence and presence of magnetic processing due to the morphological change of **Ph(4)C_{60}** nanostrucures.

1. Introduction

Applying strong magnetic fields to materials induces huge magnetic field effects (MFEs). It is expected to create highly functional nanomaterials containing new physical and chemical properties, since new interfaces or nanostructures are constructed by strong magnetic fields. [1]

Magnetic orientation of crystals [2-4], polymers [5,6], fibrin fiber [7,8], and carbon nanotubes [9-11] by strong magnetic fields due to a superconducting magnet has been widely investigated. 2D- or 3D-patterns of silver dendrites in a strong magnetic field were reported by Mogi et al. [12,13] and Katsuki et al. [14,15]. Significant morphological changes induced by magnetic fields were interpreted in terms of magnetohydrodynamic (MHD) mechanism, in which Lorentz force affects the motions of ions in a magnetic field, and/or by magnetic force which is caused by a gradient magnetic field. MFEs on growth morphology in electropolymerization, photoelectrochemical reactions, and redox behaviors of polypyrole as a conducting polymer have been reported by Mogi et al. [16-18] The MFEs were explained by the diamagnetic orientation of the polymers. Recently, Tanimoto et al. reported attractive 3D morphological chirality in membrane tubes prepared by a silicate garden reaction using strong magnetic field. [19,20]

It has also been reported that C_{60} and its derivatives formed optically transparent microscopic clusters in some mixed solvents. [21,22] The clustering behavior of C_{60} is associated with the strong hydrophobic interactions among the C_{60} units. Photoinduced electron-transfer and photoelectrochemical reactions using the C_{60} clusters have been reported because of their interesting properties. [21,22] Recently, we examined photoinduced electron-transfer reactions and MFEs on the dynamics of a radical pair that was generated from the intermolecular electron-transfer reaction between a C_{60} cluster and methyl phenothiazine (**MePH**) (Figure 1). [23] A fullerene derivative containing a positive charge (a monopyrrolidinium salt) (**$C_{60}N^+$**) (Figure 1) forms optically transparent clusters in the THF/H_2O mixed solvent. Also, transient absorption spectra of the clusters of **$C_{60}N^+$**-MePH system following laser excitation at 355 nm indicated that the photoinduced intermolecular electron-transfer from the triplet excited state of phenothiazine (PH) to **$C_{60}N^+$** cluster (**$C_{60}N^+$**)$_n$ occurred and the yield of the escaped phenothiazine cation radical ($PH^{\bullet +}$) increased with increase of magnetic field. These MFEs can be explained by spin-lattice relaxation mechanism in radical pair mechanism and most likely ascribed to the nanoclusters of **$C_{60}N^+$**.

We also reported MFEs on the photoelectrochemical reactions of photosensitive electrodes modified with nanoclusters containing of **$C_{60}N^+$** and **MePH**, intended for utilization of C_{60} as photofunctional nanodevices. [24] Photosensitive electrodes with the nanoclusters of **$C_{60}N^+$**and **MePH** were fabricated on the electrodes modified with a self-assembled monolayer of $HS(CH_2)_2SO_3^-Na^+$ on a gold substrate. The photocurrents increased with increasing an external magnetic field, typically 3 % at 0.5 T. The MFEs are ascribable to the contribution of triplet radical pair that are generated by photoinduced intermolecular electron-transfer from **MePH** to triplet excited state of nanocluster of **$C_{60}N^+$**, and can be explained in terms of spin-lattice relaxation mechanism in radical pair mechanism.

Besides the MFEs in photochemistry and photoelectrochemistry, we recently reported the effects of magnetic processing on morphological, electrochemical, and photoelectrochemical properties of electrodes modified with nanoclusters of **$C_{60}N^+$** and **MePH** in the absence and presence of magnetic processing due to strong magnetic field. [25] In the study, we used a superconducting magnet with horizontal direction of magnetic field. We demonstrated that the morphology of nanostructrures, electrochemical, and photoelectrochemical properties in the electrodes modified with nanoclusters of C_{60} are controlled by applying strong magnetic field. The morphological effects of applying magnetic fields are probably ascribed to MHD mechanism and/or the convective flow of suspension produced by magnetic force, since the **$C_{60}N^+$-MePH** clusters has positive charge.

In this study, we examined effects of magnetic processing on morphological, electrochemical, and photoelectrochemical properties of electrodes modified with nanostructures of a phenothiazine-C_{60} linked compound with four methylene group (**Ph(4)C_{60}**) in the absence and presence of magnetic processing in three different magnetic environments with hypogravity, hypergravity, and normal gravity due to strong magnetic field with vertical direction of magnetic field. In the study, the MHD mechanism due to Lorentz force is ineffective in the effects of magnetic processing, since the nanostructures consisting of **Ph(4)C$_{60}$** possess no charge.

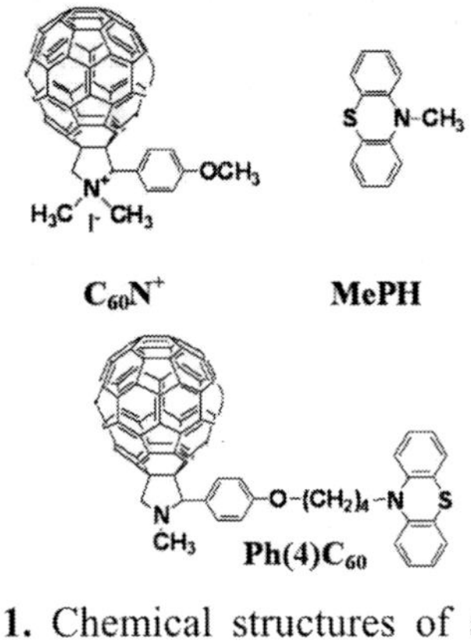

Figure 1. Chemical structures of a C_{60} derivative (**$C_{60}N^+$**), methylphenothiazine (**MePH**), and a phenothiazine-C_{60} linked compound (**Ph(4)C$_{60}$**).

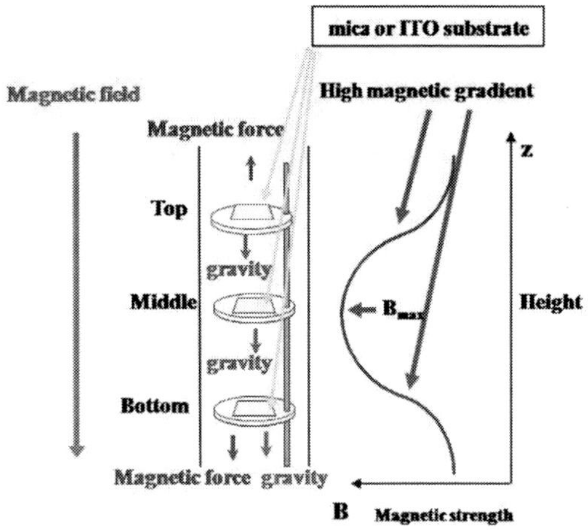

Figure 2. Experimental setup for magnetic processing.

2. Experimental

A phenothiazine-C_{60} linked compound (**Ph(4)C$_{60}$**) (Figure 1) was prepared according to the previous paper. [26] The structure of **Ph(4)C$_{60}$** was confirmed by ^{1}H-NMR and MALDI-TOF MS spectra and elemental analysis. ^{1}H-NMR spectra were measured on a JEOL model JNM-GSX 400. Mass spectra were measured by a MALDI-TOF-MS instrument (PE Biosystems Voyager DE). Steady-state absorption spectra were recorded on a Shimadzu UV-3150 spectrometer. Toluene (Dojindo, spectrosol or Wako, for Spectrochemical Anal.) and acetonitrile (CH_3CN) (Dojindo, spectrosol or Wako, for Spectrochemical Anal.) were used as received. Water was deionized with a Milli-Q system (Milipore).

Clusters of **Ph(4)C$_{60}$** were prepared by dissolving in the toluene-CH_3CN (1:1) mixed solvent due to first injection methods. [22-25] **Ph(4)C$_{60}$** formed optically transparent clusters by the following procedure. First, the toluene solution of **Ph(4)C$_{60}$** was sonicated for 15 minutes. Next, the toluene solution of **Ph(4)C$_{60}$** was injected into CH_3CN with a syringe. After sonication of the sample for 15 minutes, the solution became optically transparent in the mixed solvent (toluene:CH_3CN = 1:1) solvent.

Magnetic processing of the samples was carried out using an experimental setup as shown in Figure 2. A superconducting magnet (Japan Superconductor Technology, JMTD-LH15T40) with vertical direction of magnetic field was used as reported in the previous papers. [19,20] It has a room temperature bore tube with of 40 mm. The distribution of the magnetic field is reported in the previous papers. [19,20] The maximum field (B_{max}(vertical)) and field (B) ×gradient field (dB/dz) were 15 T and 1500 T^2/m, respectively, where z is the distance from the center of the bore tube along the tube. Three samples were placed at the positions in the bore tube, of which B and BdB/dz were 5.6 T and -940 T^2/m for the top position, 15 T and 0 T^2/m for the middle position, and 9.8 T and +1070 T^2/m for the bottom position and one was placed outside of the tube as a control.

The toluene-CH_3CN (1:1) mixed solution (50 μM) of **Ph(4)C$_{60}$** was dropped on a freshly cleaved mica or an ITO electrode (10 Ω). The samples were placed at three positions in the strong magnetic field and outside magnetic field as shown in Figure 2. Magnetic field was applied vertical to surface of the mica or the ITO electrode at room temperature. After drying the solvent at room temperature under atomspheric pressure, AFM measurements of the sample on the mica and electrochemical and photoelectrochemical measurements of the sample on the ITO electrode, so that the electrode modifed with **Ph(4)C$_{60}$** nanostructures, were carried out. In the AFM measurements, the AFM images of **Ph(4)C$_{60}$** nanostructures on the mica were recorded in tapping mode using a Nanoscope IIIa (Digital Instrument).

Electrochemical and photoelectrochemical measurements were carried out by using a three electrode cell containing of the electrode modified with **Ph(4)C$_{60}$** nanostructures as a working electrode, a platinum electrode as a counter electrodes, and a Ag/AgCl electrode as a reference electrode. Na_2SO_4 (0.1 M) was used as a supporting electrolyte according to previous paper [25].

3. Results and Discussion

*3.1. Effects of magnetic processing on morphological properties of **Ph(4)C$_{60}$** nanostructures*

The absorption spectrum of **Ph(4)C$_{60}$** in toluene is shown in Figure 3(a). The spectrum in the region of 400-800 nm is essentially identical with 6,6-closed mono-functionalized C_{60} derivative, because PH moiety in **Ph(4)C$_{60}$** has no absorption band above 400 nm. [26] For a contrast, the absorption spectrum of **Ph(4)C$_{60}$** in the toluene-CH_3CN (1:1) mixed solvent is shown in Figures 3(b) and 3(c). The absorption in the 400-800 nm regions in Figure 3(b) and 3(c) were broadened as compared with that in Figure 3(a). In addition, the spectrum changed with elapsed time after sample preparation and became same after 135 min. The peaks were slightly blue-shifted and the absorbance increased slightly with elapsed time. The results including the hyperchromic effect suggest the formation of clusters of **Ph(4)C$_{60}$** in the toluene-CH_3CN (1:1) mixed solvent as reported previously. [23-25]

The formations of clusters of **Ph(4)C$_{60}$** were also examined by AFM measurement. From the AFM image on mica placed in the absence of magnetic field at room temperature (Figure 4(a)), many small sphere-like nanoclusters were clearly observed. The results were similar to those of the clusters of **$C_{60}N^+$-MePH** in the previous papers. [24,25] The diameters of the clusters of **Ph(4)C$_{60}$** were estimated to be ca. 20 nm. However, the diameters (ca. 20 nm) of the clusters of **Ph(4)C$_{60}$** was shorter than those (ca. 100 nm) of nanoclusters of **$C_{60}N^+$-MePH** [25].

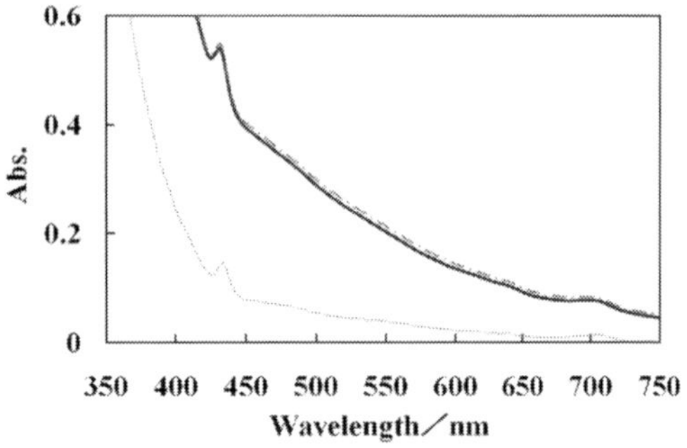

Figure 3. Absorption spectra of **Ph(4)C$_{60}$** (50 μM) in (a) toluene (green dotted line) and in toluene-CH_3CN (1:1) mixed solvent after preparation of sample at 0 min (b) (blue unbroken line) and 135 min (c) (red dash-dotted line) at room temperature.

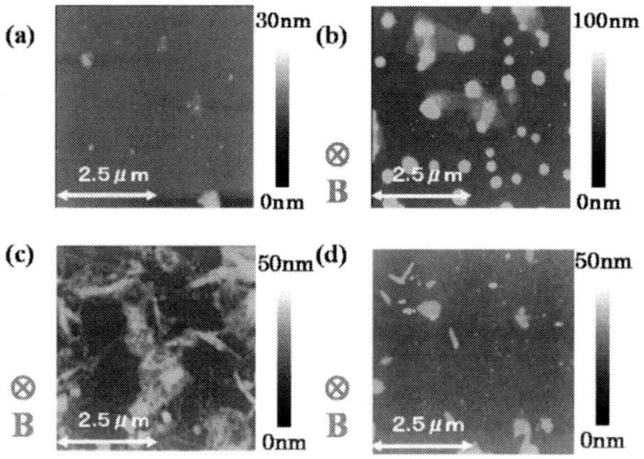

Figure 4. AFM images of **Ph(4)C$_{60}$** nanostructures obtained from **Ph(4)C$_{60}$** (50 μM) in toluene-CH_3CN (1:1) mixed solvent on mica placed in the absence (a), and the presence of magnetic processing at (b) top, (c) middle, and (d) bottom positions at room temperature.

Morphological effects of applying magnetic fields on **Ph(4)C$_{60}$** nanostructures on mica were also examined by AFM measurements. The AFM images at various magnetic enviroments at top, middle, and bottom positions are shown in Figures 4 (b)-(d). The AFM images indicated that the morphologies of **Ph(4)C$_{60}$** nanostructures in the presence of magnetic processing (Figures 4(b), (c), and (d)) were different from that in the absence of magnetic processing (Figure 4(a)). The morpologies in all AFM images were different each other (Figure 4). In the AFM images placed in the presnce of magnetic processing at middle and bottom positions, nonspherical clusters were observed (Figures 4 (c), and (d)).

At top position with hypogravity (5.6 T ; -940 T^2/m), the large spherical nanoclusters (diameter ; 60~70 nm) were observed (Figure 4(b)) and the size of the nanoclusters was much larger than those (diameter ; ca. 20 nm) in the absence of magnetic processing (Figure 4(a)). At middle position with normal gravity (15 T ; 0 T^2/m), the fiber-like nanostructure (height; ca. 20 nm) was observed. Finally, at bottom position with hypergravity (9.8 T ; +1070T^2/m), the rod-like nanostrucure (short axis :150~200nm, long axis ; 500~600 nm, height; ca. 10 nm) was observed.

The effects of magnetic processing on morphologies of **Ph(4)C_{60}** nanostructures are very interesting. In previos paper, we reported that the morphological effects of applying magnetic fields are mainly ascribed to MHD mechanism and/or the convective flow of suspension produced by magnetic force since the **$C_{60}N^+$-MePH** clusters has positive charge. [25] In the present study, the MHD mechanism due to Lorentz force is ineffective in the effects of magnetic processing (Figure 4), since **Ph(4)C_{60}** possess no charge.

Many possible mechanisms for the effects of magnetic processing on the morphologies can be considered. We suggest one of the plausible mechanisms on the basis of the results in previous papers and a a preliminary results. In previous some papers, it was reported that self-assembly of C_{60} derivative in different solvents including mixed solvents results in the formation of nanostructures. [27-29] Therefore, though we used same the mixed solvent, the interesting results might be ascribed to the different solvent properties due to the different rates of evaporation of two solvents (toluene and CH_3CN) in the toluene and CH_3CN (1:1) mixed solvent during drying process under various magnetic environments including hypogravity, normal gravity, and hypergravity at top, middle, and bottom positions. The difference of evaporation rates between toluene and CH_3CN in the absence of magnetic processing might be veried with magnetic environments at top, middle, and bottom positions.

We preliminarily examined the effects of reduced pressure on morphology of **Ph(4)C_{60}** nanostructures on a mica at the absence of magnetic processing in the preparation of sample. The AFM images obatined under reduced pressure was similar to those under atomspheric pressure in the absence of magnetic processing (Figure 4(a)). In other words, many small sphere-like nanoclusters were clearly observed similar to those in Figure 4(a). It suggested that magnetic field seems to play an important role for the difference of **Ph(4)C_{60}** nanostructures (Figure 4). The composition (toluene/CH_3CN) ratio in the mixed solvents may change by the interactions between two solvents and strong magnetic field. Further investigations are necessary to clarify the novel effects of magnetic processing on morphology of **Ph(4)C_{60}** nanostructures.

*3.2. Effects of magnetic processing on electrochemical properties of the modified electrodes with **Ph(4)C_{60}** nanostructures*

Redox potentials of ITO electrodes modified with **Ph(4)C_{60}** nanostructures were measured by cyclic voltammetry and differential pulse voltammetry measurements in the absence and presence of magnetic processing (Figures 5 and 6).

In the cyclic voltammogram (CV) in the absence of magnetic processing (Figure 5), the peak currents due to the reduction of C_{60} in the first scan were larger than those in the later scans. The peak currents due to the reduction of C_{60} were clearly observed as comparison with those due to the re-oxidation of $C_{60}\bullet^-$. The results can probably be ascribed to electron migration from $C_{60}\bullet^-$ to a neighbouring C_{60} molecule. [25] In contrast, the oxidation peaks of PH or the re-reduction peaks of $PH\bullet^+$ were not clearly observed in Figure 5. The results in CVs in the presence of magnetic processing at top, middle, and bottom positions were similar to those as shown in Figure 5.

Differential pulse voltammetry measurements of ITO electrodes modified with **Ph(4)C_{60}** nanostructures were also carried out. Differential pulse voltammograms (DPVs) for ITO electrodes modified with **Ph(4)C_{60}** nanostructures are shown in Figures 6 (a) and (b). In the case of the reduction of **Ph(4)C_{60}** nanostructures in the absence of magnetic processing, the peak due to the first reduction of C_{60} in **Ph(4)C_{60}** nanostructures was observed at -500 mV vs. Ag/AgCl (Figure 6 (a), (-1200 ← 0 mV)), while the peaks due to re-oxidation of $C_{60}\bullet^{-}$ (-508 mV vs. Ag/AgCl) and re-oxidation form $C_{60}\bullet^{2-}$ to $C_{60}\bullet^{-}$ (-902 mV vs. Ag/AgCl) were observed (Figure 6 (a), (-1200 ← 0 mV)). In contrast, in the presence of magnetic processing, the peaks due to the first reduction of C60 were negative-shifted at top (-764 mV vs. Ag/AgCl), middle (-758 mV vs. Ag/AgCl), and bottom (-600 mV vs. Ag/AgCl) positions (Figure 6 (a), (-1200 ← 0 mV)) as comparison with that (-500 mV vs. Ag/AgCl) in the absence of magnetic processing. The peaks due to re-oxidation of $C_{60}\bullet^{-}$ at top (-922mV vs. Ag/AgCl), middle (-910 mV vs. Ag/AgCl), and bottom (-902 mV vs. Ag/AgCl) positions were also negative-shifted as as comparison with that (-508 mV vs. Ag/AgCl) in the absence of magnetic processing (Figure 6 (a), (-1200 ← 0 mV)). The effects of magnetic processing on the first reduction of C_{60} of electrodes modified with **Ph(4)C$_{60}$** nanostructures was similar to those of electrodes modified with **C$_{60}$N^{+}-MePH** nanoclusters [25].

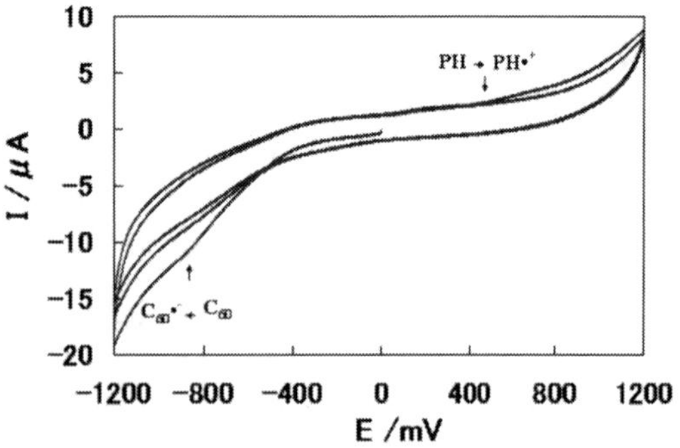

Figure 5. CV curves of **Ph(4)C$_{60}$** nanostructures on ITO electrodes in the absence of magnetic processing. (Scan direction 0→-1200→1000→-1200→1000→-1200 mV vs Ag/AgCl).

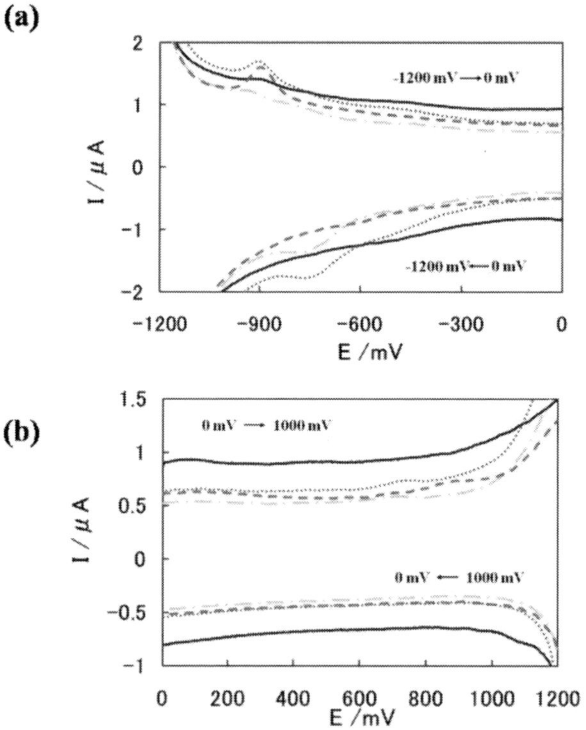

Figure 6. DPV curves of **Ph(4)C$_{60}$** nanostructures on ITO electrodes in the absence (a) (blue unbroken line) and presence of magnetic processing at (b) top (orange dash-dotted line), (c) middle (green dotted line), and (d) bottom (red broken line) positions..

On the other hand, in the case of the oxidation of **Ph(4)C$_{60}$** nanostructures, the peak due to the first oxidation peak of PH moiety was observed at 466 mV vs. Ag/AgCl in the absence of magnetic processing, while those were positive-shifted in the presence of magnetic processing at top (714 mV vs. Ag/AgCl), middle (736 mV vs. Ag/AgCl), and bottom (876 mV vs. Ag/AgCl) positions ((Figure 6 (b), (0 mV → 1000 mV)). However, the peaks due to the re-reduction of $PH\bullet^{+}$ were not appreciably observed in both the absence and presence of magnetic processing ((Figure 6 (b), (0 mV → 1000 mV)). Previously, in electrodes modified with **C$_{60}$N^{+}-MePH** mixed nanoclusters, the peak due to first oxidation of PH in **MePH** was hardly influenced by magnetic processing. [25] The effects of magnetic processing on the peak due to first oxidation of PH moiety in **Ph(4)C$_{60}$** nanostructures was similar to those on the peak due to the first reduction of C_{60} in **Ph(4)C$_{60}$** nanostructures ((Figure 6 (a)) as described above, however, was different from those on the peak due to the first oxidation of PH in **MePH** in the **$C_{60}N^+$-MePH** mixed nanoclusters [25]. The linkage between PH and C_{60} moieties in **Ph(4)C_{60}** is most likely responsible for the difference of the effects of magnetic processing in the electrodes modified with between the **Ph(4)C_{60}** nanostructures and the **$C_{60}N^+$-MePH** mixed nanoclusters.

In the previous paper, we examined electrochemical and photoelectrochemical reactions of gold electrodes modified with **$C_{60}N^+$-MePH** clusters due to electrostatic interaction between self-assembled monolayer of $HS(CH_2)_2SO_3^-Na^+$ on the gold electrode and **$C_{60}N^+$-MePH** clusters with a positive charge. [24] The two broad reduction peaks (-400 mad -800 mV vs. Ag/ACL) due to the first and second reductions of the nanocluster of **$C_{60}N^+$** were observed, while the oxidation peak (750 mV vs. Ag/ACL) due to the first oxidation of PH moiety was obtained. In addition, in the electrodes modified with **$C_{60}N^+$** clusters, the only first reduction peak due to the nanocluster of **$C_{60}N^+$** (-400 mV vs. Ag/AgCl) was observed. The reduction peaks (-400 and -800 mV vs. Ag/AgCl) of nanocluster of **$C_{60}N^+$** and the oxidation peak (750 mV vs. Ag/AgCl) of **MePH** were different of those in present modified electrode. The difference is most likely ascribed to that of the preparation of the modified electrode.

In addition, Kamat et al. reported that the two reduction peaks (-400 and -800 mV vs. Ag/AgCl) due to the first and second reductions of the C_{60} nanocluster were obtained in the CV of the C_{60} cluster film elecrophoretically deposited on the ITO electrodes. [30] Therefore, the reduction peaks in the modified electrodes in DPVs (Figure 5) are attributable to the reduction of C_{60} clusters in **Ph(4)C_{60}** nanostructures. The method of preparation of the electrodes was different from that in the present study. The electrochemical responses in the other electrodes in previous papers [24,25,30] were similar to those in the modified electrode in the present study, though the reduction potentials in the other electrodes were different from those in the modified electrode in the present study. The result may be ascribed to the difference of morphology of the nanostructures of **Ph(4)C_{60}** and/or the fabrication methods of the electrodes among the modified electrodes.

Nakashima et al. reported that the DPVs due to the reduction of C_{60} were influenced by the kind of cation electrolyte in the cast films of C_{60} lipids on electrodes. [31] The results showed the importance of the cationic charge of lipids and the microenvironments in the films for electrochemistry of C_{60}. The negative-shifts due to magnetic processing in the electrodes modified with the **$C_{60}N^+$-MePH** mixed nanoclusters can be explained by the above considerations in the previous paper [25]. Therefore, the negative-shifts of reduction of C_{60} moiety and the positive-shifts of oxidation of PH moiety in **Ph(4)C_{60}** due to magnetic processing (Figure 6(b)) are most likely explained by the same mechanism. In other words, Coulomb interaction between $C_{60}•^-$ and Na^+ cation of electrolyte, or $PH•^+$ and SO_4^{2-} anion of electrolyte are most likely responsible for the remarkable negative- or positive-shifts in the present study, since the interaction between the anion of $C_{60}•^-$ and Na^+ cation or $PH•^+$ and SO_4^{2-} anion was changed by the morphology of **Ph(4)C_{60}** nanostructures. The MFEs on the electrochemical properties of **Ph(4)C_{60}** nanostructures are in good agreement with the morphological effects of applying magnetic fields on the AFM images (Figure 4).

*3.3. Effects of magnetic processing on photoelectrochemical properties of the modified electrodes with **Ph(4)C_{60}** nanostructures.*

Photoelectrochemical properties of ITO electrodes modified with **Ph(4)C_{60}** nanostructures were also measured in the absence and presence of magnetic processing. In the absence and presence of magnetic processing, the photoirradiation of all modified electrodes afforded anodic photocurrents.

Figure 7 shows the photocurrent action spectrum of ITO electrode modified with **Ph(4)C_{60}** nanostructures in the absence magnetic processing at 0 volt vs. Ag/AgCl. The photocurrent action spectrum was in fair agreement with the absorption spectrum (Figure 3(b)) of **Ph(4)C_{60}** in the toluene-CH_3CN (1:1) mixed solvent. Similar photocurrent action spectra were observed in the presence of magnetic processing at top, middle, and bottom positions. These results indicate that the photocurrents are caused by photoexcitation of C_{60} clusters in **Ph(4)C$_{60}$** nanostructures in the absence and presence of magnetic processing. The results in the photocurrent action spectra were in fair good agreement with those in the ITO electrode modified with the **C$_{60}$N^{+}-MePH** mixed nanoclusters in previous papers [24,25].

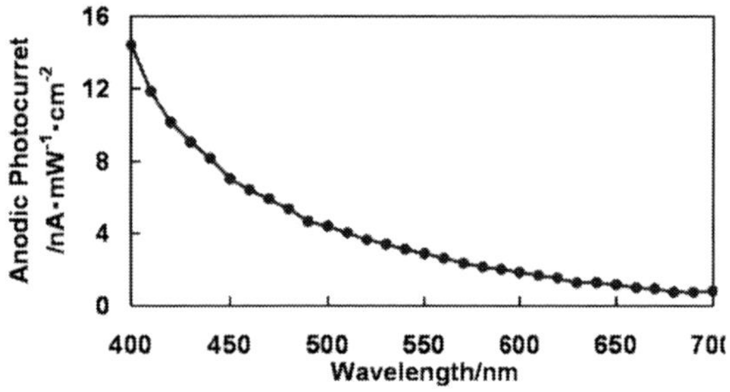

Figure 7. Photocurrent action spectrum of the electrode modified with **Ph(4)C$_{60}$** nanostructures on ITO electrodes in the absence of magnetic processing at 0 volt vs. Ag/AgCl.

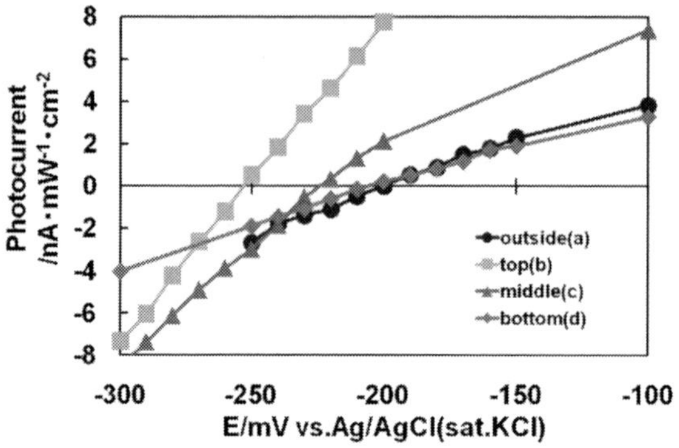

Figure 8. Potential dependencies of photocurrents of **Ph(4)C$_{60}$** nanostructures on ITO electrodes in the absence (a) (outside; ●) and presence of magnetic processing at (b) top (■), (c) middle (▲), and (d) bottom (◆) positions.

Potential dependencies of the photocurrents of the modified electrodes were examined in all ITO electrodes modified with **Ph(4)C$_{60}$** nanostructures in the absence and presence of magnetic processing (Figure 8). Potential dependencies of the photocurrents of the modified electrodes in the presence of magnetic processing at top, middle, and bottom positions (Figures 8(b)-(d)) were appreciably different from that in the absence of magnetic processing (Figure 8(a)). The potentials at zero photocurrent in the presence of magnetic processing at top position (Figure 8(b)) were remarkably negative-shifted as comparison with that in the absence of magnetic processing (Figure 8(a)). Also, the potentials at zero photocurrent in the presence of magnetic processing at middle position (Figure 8(b)) were negative-shifted as comparison with that in the absence of magnetic processing (Figure 8(a)). The magnitude of negative-shift at top position was larger than that at middle position. In contrast, the potential at zero photocurrent in the presence of magnetic processing at bottom position (Figure 8(d)) was slightly negative-shifted as comparison with that in the absence of magnetic processing (Figure 8(a)).

The reaction scheme of the photoelectrochemical reactions of the ITO electrodes modified in the absence and presence of magnetic processing is shown in Figure 9. On the basis of the reaction scheme (Figure 9), the effects of magnetic processing on the potential dependencies (Figure 8) are most likely ascribed to the difference of the first reduction potentials of C_{60} clusters in **Ph(4)C$_{60}$** nanostructures in the absence and presence of magnetic processing and are in good agreement with those in the electrochemical properties (Figures 6 (a)-(d)). The mechanism for the effects of magnetic processing on the photoelectrochemical reactions is similar to that in the previous paper [25]. The difference of the first oxidation of PH moiety in **Ph(4)C$_{60}$** nanostructures in the absence and presence of magnetic processing are probably responsible for the magnitude of photocurrent, but not for the negative-shifts of the potential at zero photocurrent.

4. Conclusion

A phenothiazine-C_{60} linked compound with four methylene group (**Ph(4)C$_{60}$**) formed optically transparent clusters (ca. 20 nm diameter) in a toluene-CH_3CN (1:1) mixed solvent via the use of first injection methods, as was verified from absorption measurements and AFM in the absence of magnetic processing. The effects of magnetic processing on the morphological, electrochemical, and photoelectrochemical properties of electrodes modified with **Ph(4)C$_{60}$** nanostructures were examined in the absence and presence of magnetic processing using three different magnetic environments including hypogravity, normal gravity, and hypergravity induced using a strong magnetic field by AFM, and electrochemical (CV and DPV) and photochemical measurements.

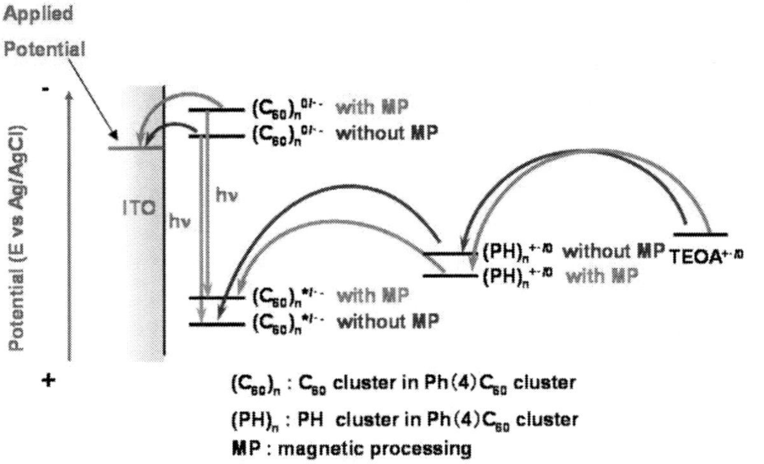

Figure 9. Schematic presentation of reaction scheme of photoelectrochemical reactions of ITO electrodes modified with **Ph(4)C$_{60}$** nanostructures in absence and presence of magnetic processing.

The AFM measurements indicated that the **Ph(4)C$_{60}$** nanostructures on mica in the absence and presence of magnetic processing at top, middle, and bottom positions were different each other. In the electrochemical measurements, the first reduction peaks corresponding to the C_{60} clusters of **Ph(4)C$_{60}$** nanostructures in the presence of magnetic processing were markedly negatively shifted in comparison with that in the absence of magnetic processing. The potential dependences of the photocurrents of the electrodes modified with **Ph(4)C$_{60}$** nanostructures in the presence of magnetic processing at top, middle, and bottom positions were also different from that in the absence of magnetic processing.

The MFEs in electrochemical and photoelectrochemical measurements can most likely be ascribed to the difference of the reduction potentials due to the C_{60} clusters between in the absence and presence of magnetic processing using the morphological change of **Ph(4)C$_{60}$** nanostructures.

We have demonstrated that the morphology of nanostructures, electrochemical, and photoelectrochemical properties in the electrodes modified with nanostructures of the Ph-C_{60} linked compound (**Ph(4)C$_{60}$**) can be controlled by applying a strong magnetic field. The present study provides useful information for designing novel nanodevices whose photofunctions can be controlled by a magnetic field. Further investigation are necessary to verify the mechanism for the effects of magnetic processing on the morphological, electrochemical, and photoelectrochemical properties of electrodes modified with **Ph(4)C$_{60}$** nanostructures. The magnetic processing using in the horizontal direction of the magnetic field and **Ph(n)C$_{60}$ (n=10)** with ten methylene group as a long spacer are now in progress.

Acknowledgment

The authors are grateful to Mr. H. Horiuchi for the preparation of photoelectrochemical cells. The present study was financially supported by the Grant-in-Aids for Scientific Research: Priority Area of "Innovative Utilization of Strong Magnetic Fields" (Area 767, No. 15085203), "Molecular Nano Dynamics" (Area 432, No. 17034051), and "Super-Hierarchical Structures" (Area 446, No. 19022027),

Scientific Research (C) (No. 17550131 and No. 18550019), and Global COE Program "Science for Future Molecular Systems" from MEXT of the Japan.

References

[1] Yamaguchi M and Tanimoto Y (eds.) 2006 Mageneto-Science Magnetic Filed Effects on Materials : Fundamentals and Applications, (Kodansha-Springer)
[2] Fujiwara M, Fukui M and Tanimoto Y 1999 *J. Phys. Chem.* B **102** 2627
[3] Fujiwara M, Tokunaga R and Tanimoto Y 1998 *J. Phys. Chem.* B **102** 5996
[4] Fujiwara M, Chidiwa T, Tokunaga R and Tanimoto Y 1998 *J. Phys. Chem.* B **108** 3417
[5] Kimira T, Yamato M, Koshimizu W, Koike M and Kawai T 2000 *Langmuir* **16** 858
[6] Kimira T, Yoshino M, Yamane T, Yamato M and Tobita M 2004 *Langmuir* **20** 5669
[7] Torbet J, Freyssinet J –M and Hudry-Clergeon G 1981 *Nature* **289** 91
[8] Yamagishi A, Takeuchi T, Higashi T and Date M 1989 *J. Phys. Soc. Jpn.* **58** 2280
[9] Fujiwara M, Oki E, Hamada M, Tanimoto Y, Mukouda I and Shimomura Y 2001 *J. Phys. Chem.* A **105** 4383
[10] Fujiwara M, Kawakami K and Tanimoto Y 2002 *Mol. Phys.* **100** 1085
[11] Yonemura H, Yamamoto Y, Yamada S, Fujiwara Y and Tanimoto Y 2008 *Sci. Tech. Adv. Mater.* **9** 024213
[12] Mogi I, Okubo S and Nakagawa Y 1991 *J. Phys. Soc. Jpn.* **60** 3200
[13] Mogi I, Kamiko M and Okubo S 1995 *Physica* B **211** 319
[14] Katsuki A, Uechi I and Tanimoto Y 2004 *Bull. Chem. Soc. Jpn.* **77** 275
[15] Tanimoto Y, Katsuki A, Yano H and Watanabe S 1997 *J. Phys. Chem.* A **101** 7359
[16] Mogi I 1996 *Bull. Chem. Soc. Jpn.* **69** 2661
[17] Mogi I, Watanabe K and Motooka M 1999 *Electrochemistry* **67** 1051
[18] Konno A, Mogi I and Watanabe K 2001 *J. Electroanal. Chem.* **507** 202
[19] Uechi I, Katsuki A, Dunin-Barkovskiy L and Tanimoto Y 2004 *J. Phys. Chem.* B **108** 2527
[20] Duan W, Kitamura S, Uechi I, Katsuki A and Tanimoto Y 2005 *J. Phys. Chem.* B **109** 13445
[21] Kamat P V, Barazzouk S, Hotchandani S and Thomas K G 2000 *Chem. Eur. J.* **6** 21
[22] Biji V, Barazzouk S, Thomas K G, George M V and Kamat P V 2001 *Langmuir* **17** 2930
[23] Yonemura H, Kuroda N, Moribe S and Yamada S 2006 *C. R. Chimie* **9** 254
[24] Yonemura H, Kuroda N and Yamada S 2006 *Sci. Tech. Adv. Mater.* **7** 643
[25] Yonemura H, Wakita Y, Kuroda N, Yamada S, Fujiwara Y and Tanimoto Y 2008 *Jpn. J. Appl. Phys.* **47** 1178
[26] Moribe S, Yonemura H and Yamada S 2007 *Chem. Phys.* **334** 242
[27] Nakanishi T, Schmitt W, Michinobu T, Kurth D G and Ariga K 2005 *Chem.Commun.* 5982
[28] Georgakilas V, Pellarini F, Prato M, Guldi D M, Melle-Franco M and Zerbetto F 2002 *Proc. Natl. Acad. Sci. USA* **99** 5075
[29] Guldi M, Zerbetto F, Georgakilas V and Prato M 2005 *Acc Chem Res.* **38** 38
[30] Barazzouk S, Hotchandani S and Kamat P D 2001 *Adv. Mater.* **13** 1614
[31] Nakanishi T, Ohwaki H, Tanaka H, Murakami H, Sagara T and Nakashima N 2004 *J. Phys. Chem.* B **108** 7754

Chiral electrode behavior of magneto-electrodeposited Cu-Cu_2O films

I Mogi and K Watanabe

Institute for Materials Research, Tohoku University, Sendai 980-8577, Japan

E-mail: mogi@imr.tohoku.ac.jp

Abstract. Chiral properties were investigated in Cu and Cu_2O composite films prepared by magneto-electrodeposition. The Cu-Cu_2O films were electrodeposited in a $CuSO_4$ and Na_2SO_4 aqueous solution under magnetic fields *B* (2 and 5 T) parallel (+*B*) or antiparallel (–*B*) to the faradaic current. The magneto-electrodeposited films were used as modified electrodes, and the chiral voltammetric behaviors were observed as follows. The 2T-film electrodes of Cu-Cu_2O(15%) showed the oxidation current difference between D- and L-glucoses. The voltammograms of L-DOPA (3-(3,4-dihydroxyphenyl)alanine) showed the peak potential difference between the +5T-film and –5T-film electrodes of Cu-Cu_2O(40%).

1. Introduction

Recent development of magnetoscience has provided a new strategy for material processing, namely, magneto-processing of materials. Combinations of magnetic fields and conventional processes allow control of mass transfer, molecular orientation, levitation, crystal morphology, etc., however, there remains an intriguing and open issue, that is, magnetic control of chirality. Rikken and Raupach exploited magnetochiral chemistry in a photochemical reaction. They found an enantioselective magnetic field effect in the photo-induced dissociation of the chiral tris(oxalato)-Cr(III) complex and obtained the enantiomeric excess of 0.01% under a magnetic field of 10 T [1].

We have explored chirality in magneto-electrochemical processing. When a magnetic field is imposed to an electrochemical cell, the Lorentz force acting on the faradaic current causes convection of the electrolytic solution. This is well known as the magnetohydrodynamic (MHD) effect [2,3]. The Lorentz force induces spiral structures in the two-dimensional growth of magneto-electrodeposited metals [4-7] and conducting polymers [8], and the three-dimensional helical structures in silicate membrane growth [9].

Chiral structures can be used for the enantioselective recognition of chiral molecules. If the chiral structures can be formed in nanometer scales on the magneto-electrodeposited films, such film surfaces would serve as an enantioselective catalyst. As the first attempt, we tried to prepare chiral conducting polymer films by magneto-electropolymerization and found that such polyaniline films exhibited chiral electrode behavior for the oxidation of ascorbic acid [10,11]. This result suggested that the Lorentz force affected the growth probabilities of right- and left-handed helical polyaniline chains.

The chiral electrode behavior was also found in magneto-electrodeposited Ag films, which exhibited different oxidation currents between D- and L-glucoses [12]. Aogaki pointed out that the micro-MHD effect plays an important role in electrodeposition under the magnetic fields parallel to the faradaic currents [13]. In electrodeposition processes, non-equilibrium fluctuation produces a lot of humps on the deposit surfaces. The faradaic currents around such humps are not parallel to the magnetic field, so that the Lorentz force could cause vortex-like convections in the local areas around the humps. Such local convections are expected to induce chiral structures on the deposit surfaces. The chirality is thus considered to be an intrinsic feature in magneto-electrodeposition. To confirm this concept, it is necessary to explore various kinds of magneto-electrodeposited films. In this paper, we report the magneto-electrodeposition of Cu and Cu_2O composite films and their chiral electrode behaviors.

2. Experimental

All chemicals were reagent grade and were used as received. All aqueous solutions were prepared with distilled water. For electrochemical experiments, a conventional system with the following three electrodes was employed: a platinum disc working electrode with a diameter of 3 mm, a platinum plate counter electrode, a Ag | AgCl | NaCl(sat) reference electrode. A potentiostat (Princeton Applied Research model 263A) was used for all the electrochemical experiments.

Figure 1 shows a voltammogram of Cu^{2+} in a 50 mM (M = mol dm^{-3}) $CuSO_4$ aqueous solution containing 0.1 M Na_2SO_4 as a supporting electrolyte. In a small peak at 67 mV, Cu^{2+} is reduced to Cu(I), forming cuprous oxide Cu_2O electrodeposits [14]. In a large peak at −70 mV, Cu^{2+} is reduced to Cu(0), forming Cu electrodeposits. However, Shang et al. reported that the electrodeposition in neutral solutions with pH ~ 7 leads to simultaneous deposition of Cu and Cu_2O in a wide potential range [15]. In our experiment, electrodeposition of Cu and Cu_2O composite films was conducted at 0 and 40 mV (vs Ag | AgCl) by a passing charge of 0.15 C cm^{-2} in a 50 mM $CuSO_4$ and 0.1 M Na_2SO_4 aqueous solution. The molar ratios of Cu and Cu_2O in the composite films depended on the deposition potential and were determined electrochemically, as mentioned in the next section.

The electrochemical cell was placed at the bore center in a cryocooled superconducting magnet (Sumitomo Heavy Industries Ltd.), which can produce magnetic fields of up to 5 T. An imposed magnetic field B was perpendicular to the working electrode surface, and it was parallel ($+B$) or antiparallel ($-B$) to faradaic currents. The electrodeposited films prepared at +2 T or −2 T, for example, are called the +2T-film or the −2T-film, respectively. The temperature within the magnet bore was controlled at 25°C by circulating thermo-regulated water.

The magneto-electrodeposited films were used as modified electrodes, and their chiral properties were examined by measuring the voltammograms of 20 mM D- and, L-glucoses in 0.1 M NaOH

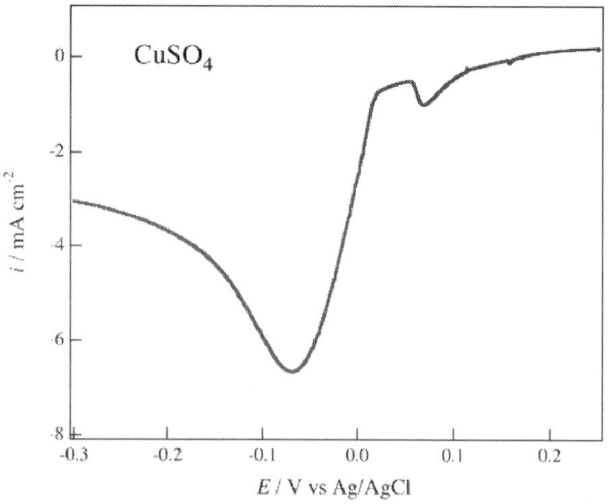

Figure 1. Voltammograms of 50 mM $CuSO_4$ in a 0.1 M Na_2SO_4 aqueous solution. The potential sweep rate was 10 mV s^{-1}.

aqueous solutions and 20 mM L-3-(3,4-dihydroxyphenyl)alanine (L-DOPA) in a phosphate buffer solution with pH = 6.86. Before the voltammetric measurements, the Cu-Cu_2O film electrodes experienced the potential sweeps in the supporting electrolyte solution as a pretreatment, and the background currents were measured for each electrode. The voltammograms of glucose and L-DOPA were measured with a potential sweep rate of 50 mV s^{-1} in the absence of a magnetic field, and the background corrections were made on each voltammogram.

3. Results and discussion

Figure 2 shows the current-time curves during the Cu-Cu_2O electrodeposition at 0 mV in 0 T and +2 T. The curve at –2 T was quite similar to that at +2 T. The magnetic field increases the electrodeposition rate after 20 sec, even though the applied field was perpendicular to the electrode surface. When the magnetic field is perpendicular to the electrode surface, the MHD effect is the minimum, but it never disappears because the current is not parallel at the edge of the electrode and the fluctuated deposit surface (the micro-MHD effect).

Figure 3 shows the voltammograms of the Cu-Cu_2O 0T-film prepared at 0 mV. The voltammograms were measured several times successively in a 0.1 M NaOH aqueous solution, and the first and the second sweep voltammograms are shown in Figure 3. In the first sweep voltammogram, the oxidation from Cu to Cu(I) occurs around a sharp peak at 0 V, and the oxidation from Cu(I) to Cu(II) occurs around the next peak at 0.2 V [16]. The oxidation current becomes very small in the second sweep voltammogram, implying that the film surface is covered by cupric oxide CuO after the first sweep.

The molar ratios of Cu and Cu_2O in the composite films were determined from the measurements of the passing charges during the film oxidation from Cu to Cu(I) in the NaOH solution. The estimated molar ratios of Cu : Cu_2O are 0.85 : 0.15 for the film prepared at 0 mV, and 0.60 : 0.40 for the film prepared at 40 mV. These ratios were nearly the same as those of the 2T- and 5T-films.

The chiral electrode behaviors of the Cu-Cu_2O(15%) films were examined for the oxidation reaction of 20 mM D- and L-glucoses in a 0.1 M NaOH aqueous solution. In the alkaline solution, CuO covers the surface of the Cu electrode after the pretreatment potential sweeps, and the oxide induces electrocatalytic reactions [16]. Generally, an electrocatalytic reaction includes a specific adsorption process, which are sensitive to the surface structure of the electrode. The voltammograms of D- and L-glucoses were nearly coincident each other on the 0T-film electrode.

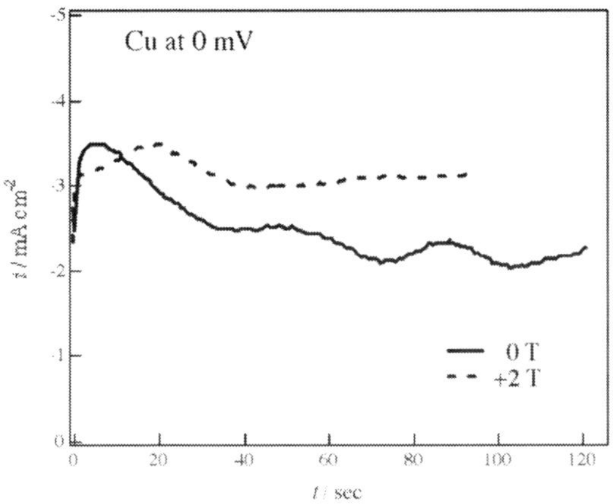

Figure 2. Current-time curves during the electrodeposition of Cu-Cu_2O at 0 mV in 0 T and +2 T.

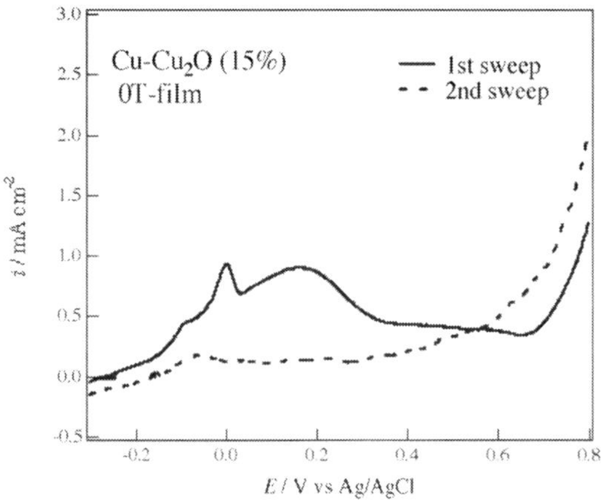

Figure 3. Voltammograms of the Cu-Cu_2O(15%) 0T-film in a 0.1 M NaOH aqueous solution. The potential sweep rate was 50 mV s^{-1}.

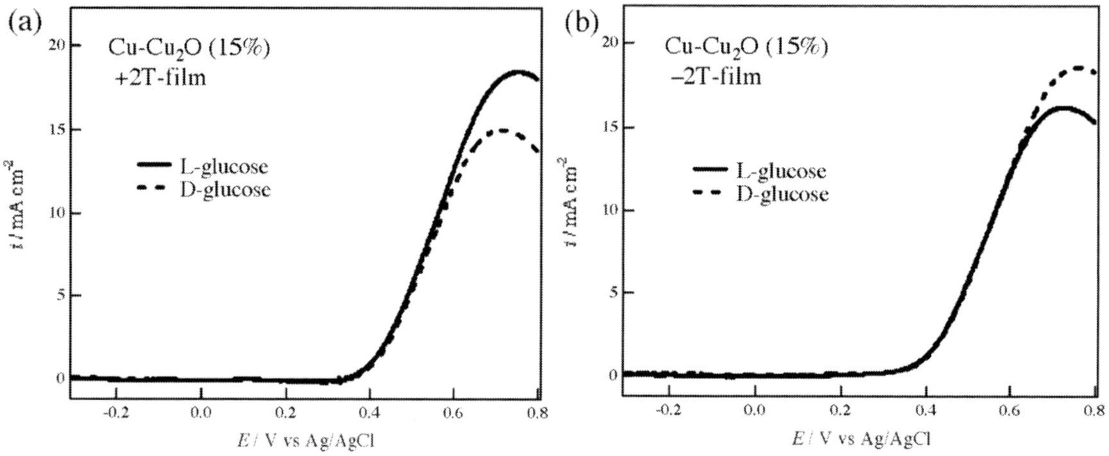

Figure 4. Voltammograms of 20 mM D- and L-glucoses in a 0.1 M NaOH aqueous solution on the Cu-Cu_2O(15%) electrodes of (a) the +2T-film and (b) the –2T-film. The potential sweep rate was 50 mV s^{-1}.

Figure 4 shows voltammograms of D- and L-glucoses on the +2T-film and –2T-film electrodes of Cu-Cu_2O(15%). The oxidation peak of glucose appears around 0.7 V. On the +2T-film electrode, the oxidation current of L-glucose is larger than that of D-glucose (Figure 4(a)). On the contrary, the current of D-glucose is larger than that of L-glucose on the –2T-film electrode (Figure 4(b)). These facts indicate that the magneto-electrodeposition induces chirality in the Cu-Cu_2O(15%) films and that the origin of this chirality is the Lorentz force because other magnetic field effects are independent of the polarity of magnetic fields. The 2T-film electrodes of Cu-Cu_2O(15%) possess the ability of chiral recognition for glucoses.

The chiral electrode behaviors of the magneto-electrodeposited Cu-Cu_2O(40%) films were examined for electrochemical reactions of an amino acid. L-DOPA is a kind of an amino acid with a redox functional group of dihydroxyphenyl, as shown in Figure 5. The cyclic voltammograms of L-DOPA were measured on the +5T- film and –5T-film electrodes in a phosphate buffer solution with pH = 6.86. The result is shown in Figure 6(a). The oxidation of L-DOPA occurs around 0.3 ~ 0.4 V on the Cu-Cu_2O film electrode. It is noted that the peak potentials are different between the +5T- film and –5T-film electrodes; 351 mV on the +5T-film electrode and 366 mV on the –5T-film electrode. Figure 6(b) shows cyclic voltammograms of the racemic mixtures of D- and L-DOPA on the +5T- film and –5T-film electrodes. The peak potentials of the racemic mixture are coincident each other between the two electrodes. These results indicate that the +5T-film and –5T-film electrodes of Cu-Cu_2O(40%) have chirality and possess the ability of chiral recognition of L-DOPA.

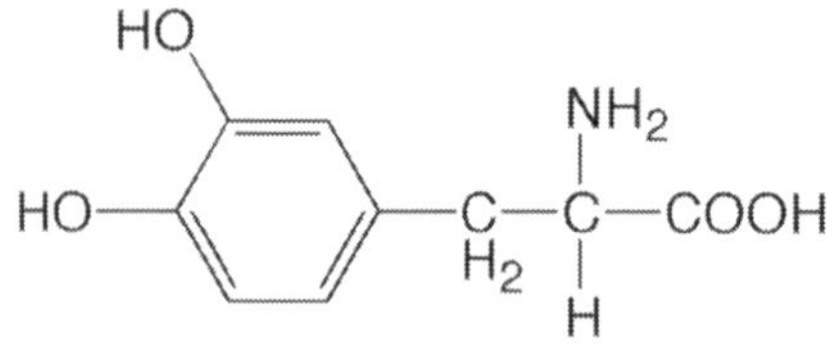

L-3-(3,4-Dihydroxyphenyl)alanine
(L-DOPA)

Figure 5. Molecular structure of L-DOPA

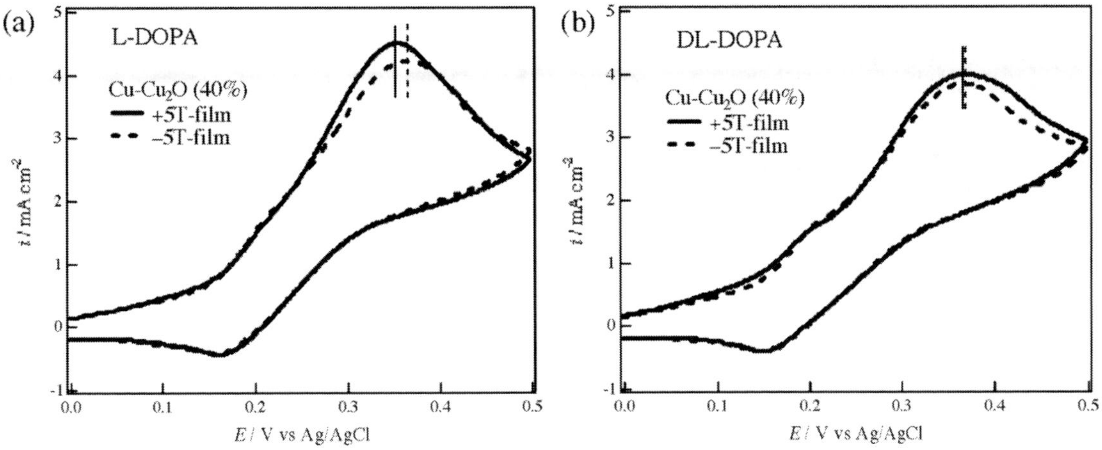

Figure 6. Cyclic voltammograms of 20 mM (a) L-DOPA and (b) the racemic mixture of L- and D-DOPA in a phosphate buffer solution with pH = 6.86 on the +5T-film and the –5T-film electrodes of Cu-Cu_2O(40%). The potential sweep rate was 50 mV s^{-1}. The peak positions are indicated by vertical lines.

It is known that the chiral voltammetric behavior depends on the electrocataliticity of an electrode reaction. Kashiwagi et al. reported the electrocatalytic voltammograms of R- and S-1-phenylethanols on modified electrodes with a chiral self-assembled monolayer, and showed that the chiral behavior appears in the electrocatalytic currents [17]. On the other hands, if the electrocataliticity is poor, the chiral behavior appears on the peak potentials. Switzer reported that the voltammograms of L- and D-tartaric acids have different peak potentials on chiral CuO electrodes [18]. It seems that the former corresponds to the result of Figure 4 and the latter corresponds to the result of Figure 6. Similar results were observed in the electrocatalytic oxidation of ascorbic acid on magneto-electropolymerized polyaniline film electrodes [11].

4. Conclusion

We have demonstrated that the magneto-electrodeposition induces chirality in the Cu-Cu_2O composite films. The 2T-film electrodes of Cu-Cu_2O(15%) exhibit the oxidation current difference between D- and L-glucoses. The voltammograms of L-DOPA showed the peak potential difference between the +5T-film and –5T-film electrodes of Cu-Cu_2O(40%). These results indicate that the magneto-electrodeposited Cu-Cu_2O films possess the ability of chiral recognition for glucoses and DOPA.

Acknowledgement
The magneto-electrodeposition experiments were performed in the High Field Laboratory for Superconducting Materials, IMR Tohoku University.

References

[1] Rikken G L J A and Raupach E 2000 *Nature* **405** 932
[2] Aogaki R, Fueki K and Mukaibo T 1975 *Denki Kagaku* **43** 504
[3] Fahidy T Z 1983 *J. Appl. Electrochem.* **13** 553
[4] Mogi I, Okubo S and Nakagawa Y 1991 *J. Phys. Soc. Jpn.* **60** 3200
[5] Mogi I, Kamiko M and Okubo S 1995 *Physica B* **211** 319
[6] Coey J M D, Hinds G and Lyons M E G 1999 *Europhys. Lett.* **47** 267
[7] Heresanu V, Ballou R and Molho P 2003 *Magnetohydrodynamics* **39** 461
[8] Mogi I and Kamiko M 1996 *Denki Kagaku* **64** 842
[9] Uechi I, Katsuki A, Dunin-Barkovskiy L and Tanimoto Y 2004 *J. Phys. Chem. B* **108** 2527

[10] Mogi I and Watanabe K 2005 Jpn. *J. Appl. Phys.* **44** L199
[11] Mogi I and Watanabe K 2007 *J. Solid State Electrochem.* **11** 751
[12] Mogi I and Watanabe K 2007 *ISIJ Int.* **47** 585
[13] Aogaki R 2003 *Magnetohydrodynamics* **39** 453
[14] Wijesundera R P, Hidaka M, Koga K, Sakai M and Siripala W 2006 *Thin Solid Films* **500** 241
[15] Shang W, Shi X, Zhang X, Ma C and Wang C 2007 *Appl. Phys. A* **87** 129
[16] Torto N, Ruzgas T and Gorton L 1999 *J. Electroanal. Chem.* **464** 252
[17] Kashiwagi Y, Uchiyama K, Kurashima F, Anzai J and Osa T 1999 *Electrochemistry* **67** 900
[18] Switzer J A 2004 *Interface* **13** 34

Magnetic Field Increases the Surface Tension of Water

Y Fujimura and M Iino

Department of Life and Environmental Sciences, Faculty of Engineering,

Chiba Institute of Technology, Tsudanuma, Narashino-shi, Chiba 275-0016, Japan

Email: g0473028NL@it-chiba.ac.jp, masaaki.iino@it-chiba.ac.jp

Abstract. We studied what effects magnetic fields have on the surface tension of water-air interfaces. We measured the surface tension by means of the surface wave resonance method with very high precision. The surface tension increased by 1.83 ± 0.18 % at the magnetic field of 10 T. As for artificial effects and possible contributions to the surface tension increase, it seems most likely that the stabilization of hydrogen bonds increases the bulk Helmholtz's free energy, at least at the surface, which thereby increases the surface tension.

1. Introduction

Recently, effects of strong magnetic fields on the physical properties of water such as: infrared spectrum [1], refractive index [2], viscosity [3], and melting temperature [4] have been reported. These effects were attributed to the magnetic field stabilizing hydrogen bonds of water. However, the increase in Gibbs's free energy due to the stabilization of hydrogen bonds would generate a loss by the external work (pressure-volume term) due to an enlargement of the specific volume by hydrogen bonds. Therefore, the net increase in the Gibbs's free energy must be equal to that of the Gibbs's free energy observed as the surface level difference of ~4 cm at 10 T (so called "Moses effect"). This is comparable to the diamagnetic energy of ~6.5 mJ/mol (or 3×10^{-8}) at 10 T. Therefore, the magnetic field would not affect the bulk characteristics and the phase equilibrium. On the other hand, the surface tension of a flat interface is equal to the Helmholtz's free energy per unit area of the interface, which is given by the internal energy per unit area of the interface, minus the entropy per unit area of the interface, and multiplied by the temperature. That is to say, the increase in the energy of hydrogen bonds directly reflects the increase in the surface tension. Since the biological systems include water-air or water-macromolecule interfaces, a biological system would relate to the surface thermodynamic properties, including the surface tension of water. Langmuir-Blodgett (LB) membrane technology also strongly relates to the surface tension. Therefore, clarification of the magnetic field effect on the surface tension of water is desired in both science as well as in technology.

2. Method

Ultra-pure water (Lot. DPF2293, Wako, Osaka, Japan) was used for the sample without further purification. We have constructed a system for measuring the surface tension of water under the magnetic field by means of the surface wave resonance method [3]. We used a rectangular optical cell ($30 \times 10 \times 45$; made of fused quartz) as the surface wave resonator. Interleaved electrodes were fixed outside of the cell as a generator and a detector. A voltage controlled oscillator (FG-273A, Kenwood

TMI Co., Tokyo, Japan) generated an alternative voltage temporal wave. Surface waves were excited by applying the alternative voltage to the electrode of the generator. Then, standing waves were detected as capacitance changes of the electrodes of the detector. The capacitance changes were than converted to a frequency modulated high frequency signal by a capacitance to frequency converter (a clap oscillator, hand made). The high frequency signal was detected by a frequency modulation receiver (AR5000, AOR, Ltd., Tokyo, Japan), and led to a lock-in amp (LI-575, NF Co., Yokohama, Japan) with 2f mode. The quadrature output of the lock-in amp led to the control voltage of the voltage controlled oscillator through a control circuit (hand made). The resonator was set in the field center of the superconducting magnet (100 mm in bore, JMTD-10T100MK, JASTEC, Koube, Hyougo, Japan). The water-air interface was perpendicular to the magnetic field. The experimental system is shown in Fig. 1. The measurement system formed a feedback loop to lock onto a surface wave resonance. We chose the modes 4 and 5, which are shown in Fig. 2, with the in-phase and quadrature outputs from the lock-in amplifier. Zero crossing points of the quadrature output are locked by the phase locked loop system mentioned earlier. All measurements were carried out in a temperature-stabilized cupper plate at 298.2±0.1 K which was stabilized by a water bath (LTB-125, As One Co., Osaka, Japan). Figure 3 shows the stability of a locked resonance frequency in mode 5. The interfacial tension between water and air was obtained from the dispersion of surface waves given by [5-7]

$$\omega^2 = \frac{(\rho_l - \rho_g) g k + \sigma k^3}{\rho_l \coth(kd_l) + \rho_g \coth(kd_g)} , \qquad (1)$$

where ω is the angular frequency of the surface waves, g is the acceleration due to gravity, σ is the interfacial tension, k is the wave number, ρ_l and ρ_g are the mass densities of the water and the air, respectively, and d_l and d_g are the thicknesses of water and air, respectively. To prevent the sample from heating up as well as any nonlinear effect on Eq. (1), the surface wave amplitude was kept under 1 μm.

With solving simultaneously the dispersion equations of several resonance modes to cancel the inhomogeneity of the magnetic field, we were able to obtain measurements of the surface tension very precisely.

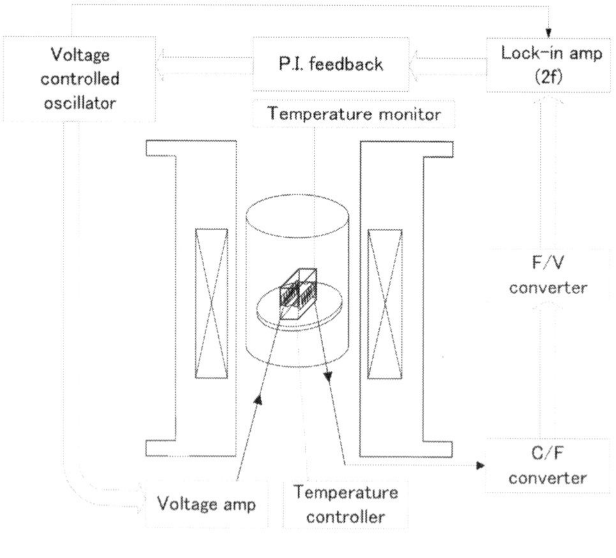

Figure 1. The experimental system.

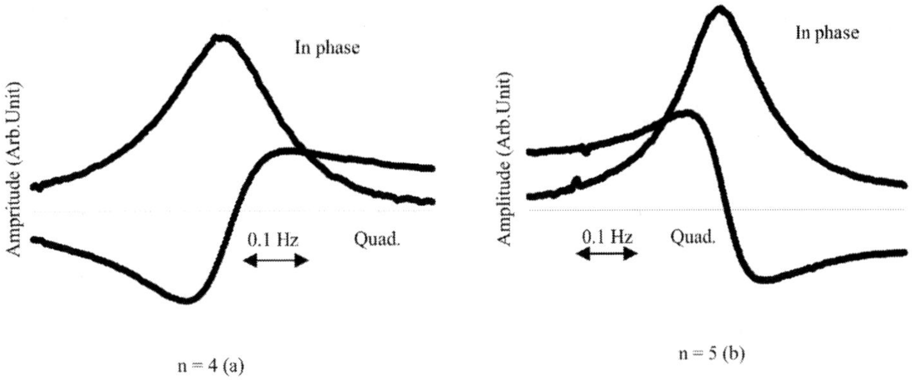

Figure 2. In-phase and quadrature outputs from the lock-in amplifier: mode 4 (a) and mode 5 (b).

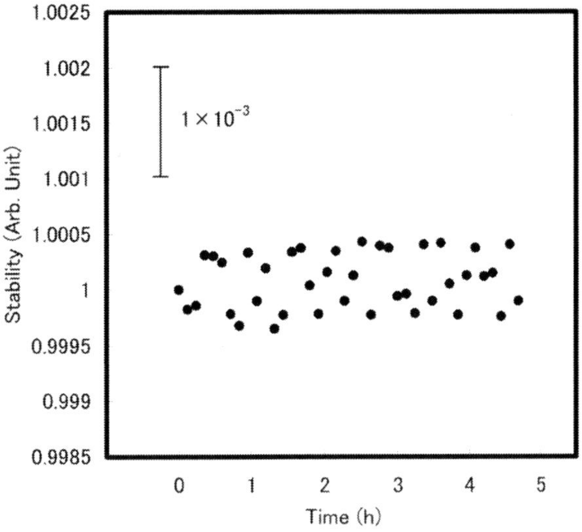

Figure 3. Stability of a resonance frequency.

3. Result and Discussion

We calculated the surface tension from the resonance frequencies. The values of the surface tension were 71.96 ± 0.14 mN/m at 0 T and 73.31 ± 0.16 mN/m at 10 T. The surface tension increased by 1.32 ±0.13 mN/m or 1.83±0.18 % at 10 T. Figure 4 shows the effect of the magnetic field on the measured surface tension over time. Initially, the value of the surface tension was 71.96 mN/m. The surface tension increased in response to the application of the magnetic field, and reached a value of 73.31 mN/m. With removing the magnetic field, the surface tension decreased and returned to the initial value.

At first we considered the artificial errors due to temperature fluctuations and the inhomogeneity of the magnetic field. Given that the temperature derivative of the surface tension of water was ~0.16

$mNm^{-1}K^{-1}$ at 298 K, a temperature fluctuation of ±0.1 K yielded a surface tension error of ±16 μN/m or ±0.02 % as shown in Fig. 5. The observed surface tension increases sufficiently exceeded the temperature fluctuations and the systematic errors of the thermometer.

The inhomogeneity of the magnetic field enlarged the length along the surface, which decreased the value of *k* and lowered the resonance frequency and apparent value of the surface tension. In our calculation of Eq. (1), the elongation of the interface was 17 ppm at 10 T, which yielded 0.2 % of the error in the value of the surface tension if a single mode was used to solve Eq. (1). The effect of the in homogeneity of the magnetic field was canceled by the above-mentioned two mode simultaneous analysis to be roughly 0.05 % as shown in Fig. 5.

Next, we discuss the additional effects of oxygen, diamagnetic energy, and hydrogen bond stabilization on the surface tension increase. In general, molecules localized (delocalized) at the surface decrease (increase) the surface tension of a liquid because of the surface pressure of the molecules at the surface. The concentration of oxygen in the air is much greater than in water, indicating that the oxygen decreases the surface tension of water. If the magnetic field removes the decrease in the surface tension due to oxygen, the resulting increase in the surface tension is about 6 orders of magnitude less than the observed increase in the surface tension at 10 T as shown in Fig. 5.

The diamagnetic energy of water due to the magnetic field is only −6.5 mJ/mol at 10 T, which is comparable to the increase in the Gibbs's free energy, and only 0.3-1.2 ppm of the hydrogen bonding energy 5.4-18.8 kJ/mol [8] as shown in Fig. 5.

If the magnetic field stabilizes the hydrogen bonds of water, the increase in the diamagnetic energy affects not only the internal energy of water, but also the entropy of water, resulting in a large increase in the bulk Helmholtz's free energy. Iwasaka et al. and Hosoda et al. found that the hydrogen bond became more stable under the magnetic field. Their data implied that the hydrogen bonds are stabilized by roughly 1% at 10 T. This subsequently yields an increase in the Helmholtz's free energy of the order of 1 J/mol at 10 T, if we assume that the Helmholtz's free energy of water equals that which is due to the hydrogen bond formation. An increase of ~1J/mol in the bulk Helmholtz's free energy would explain the magnetic increase in the surface tension as shown in Fig. 5.

The increase in the interfacial tension implies that the magnetic fields strengthen the hydrophobic bonds in biological systems and LB membrane, and steric conformation of biological molecules such as enzymes.

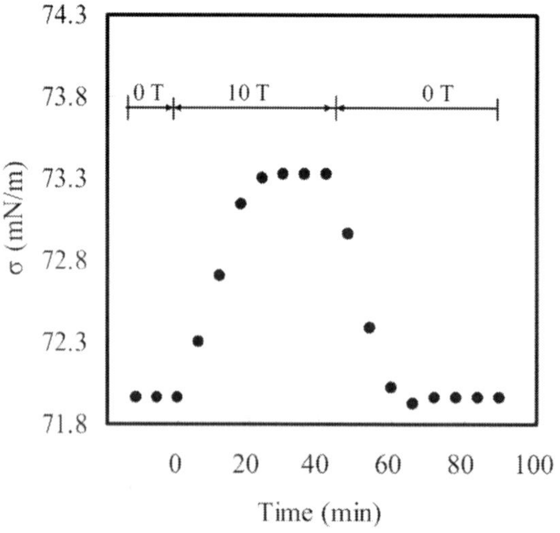

Figure 4. Effect of 10 T magnetic field on the measured surface tension over time.

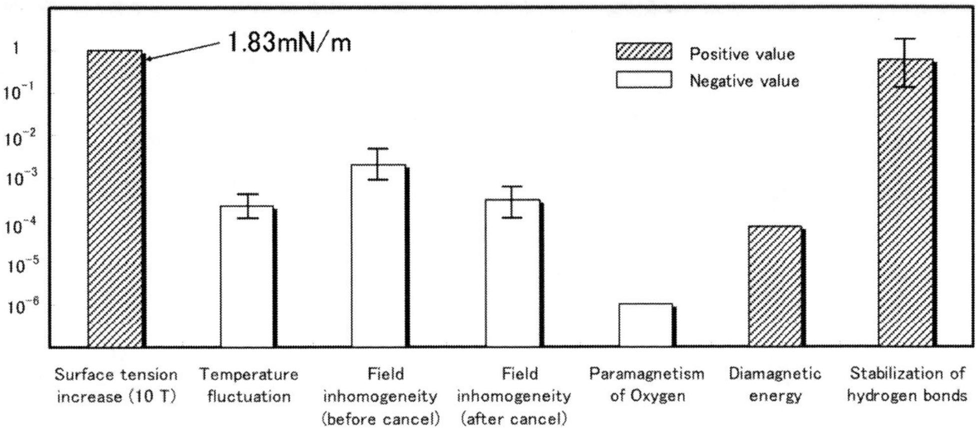

Figure 5. Factors suggesting an apparent increase of the surface tension.

Acknowledgments
The authors would like to thank Dr. Sogoshi of Saitama University for many useful discussions. This work was partly supported by Grant-In-Aid for High Technology Research from the Japanese Ministry of Education, Science, Sport, and Culture.

References
[1] Iwasaka M and Ueno S 1998 *J. Appl. Phys.* **83** 6459
[2] Hosoda H, Mori H, Sogoshi N, Nagasawa A and Nakabayashi S 2004 *J. Phys. Chem.* **108** 1641
[3] Ghauri S A and Ansari M S 2006 *J. Appl. Phys.* **100** 066101
[4] Inaba H, Saitou T, Tozaki K and Hayashi H 2004 *J. Appl. Phys.* **96** 6127
[5] Iino M, Suzuki M, Ikushima A J and Okuda Y 1985 *J. Low Temp. Phys.* **59** 291
[6] Iino M, Suzuki M and Ikushima A J 1986 *J. Low Temp. Phys.* **63** 495
[7] Iino M, Suzuki M, Ikushima A J and Okuda Y 1983 *Jpn. J. Appl. Phys.* **23-1** 54
[8] Eisenberg D and Kauzmann W 1969 *The Structure and the Properties of Water* (Oxford University Press, Oxford)

Effects of Gradient Magnetic Fields on CO_2 Sublimation in Dry Ice

M Iwasaka

Department of Medical System Engineering, Graduate School of Engineering, Chiba University, Chiba 263-8522, Japan

E-mail: iwasaka@faculty.chiba-u.jp & iwasaka-m@umin.ac.jp

Abstract. A process of CO_2 sublimation in a fragment of dry ice was observed under static gradient magnetic fields of up to 7 Tesla. Results indicated that Tesla order magnetic fields enhanced the release of CO_2 gas from solid matter containing carbon dioxide. Paramagnetic oxygen molecules transported the sublimated particles containing carbon dioxide to a higher magnetic field.

1. Introduction

Magnetic fields in the range of micro-Tesla to milli-Tesla are unique properties of our planet and civilization. Mars has no magnetic fields and does have an abundant supply of dry ice at both of its poles. In contrast, as a gaseous planet, Jupiter, which has strong internal magnetic moments, has magnetic fields about ten times greater than those of Earth. Recently, our planet has begun to experience an increase in the volume of atmospheric CO_2 gases, which is thought to be life-threatening if it continues. Many reasons for the increased generation of CO_2 gases have been investigated and discussed with the aim of preventing or even reversing their release.

On the other hand, many studies on magnetic separation technique were reported concerning gaseous, liquid, and solid-state materials in gradient magnetic fields [1-7]. Most of them focused on the importance of the difference in magnetic susceptibilities of molecules which were involved in an experimental system.

In the present study, we observe the sublimation of dry ice into the atmosphere in the presence and absence of a strong static magnetic field, which is generated by a superconducting magnet. The results show that magnetic fields on the order of a Tesla can possibly enhance the release of CO_2 gas from dry ice. A hypothesis concerning the passive transportation of CO_2 by magnetic force acting on oxygen is proposed and discussed based on a qualitative experiment.

2. Methods

2.1. Magnetic field generators

We utilized superconducting magnets with a room temperature bore as magnetic field generators. The utilized 5-T vertical superconducting magnet (Oxford Inst., Microstat BT) generated a maximum field at the center and provided gradient fields of 4T with 50T/m at 20 mm from the center. In case with a horizontal superconducting magnet of maximum 8 Tesla (Oxford Inst.), a gradient field of 7T with 40T/m was utilized.

2.2. Macroscopic observation of CO_2 sublimation and gas flow control

CO_2 sublimation in a fragment of dry ice, which was set in the bore of magnet, was monitored both by a CCD video camera and directly by eyes. A commercial dry ice was used for the experiments. Dry ice fragment with an edge about 20 mm in length was chosen for the experiments. In the experiment in 5-T vertical bore of the superconducting magnet, a gas flow condition of the atmosphere near the dry ice was roughly controlled by a tubing outlet which provided a gas from either a gas bottle of He, O_2, and CO_2.

In the experiment to measure a magnetic field dependence of sublimation rate in dry ice, the initial weight of dry ice fragments were set to 6.7 to 6.9 g, and the weight of fragment was measured after 8-min of magnetic field exposure at 25°C.

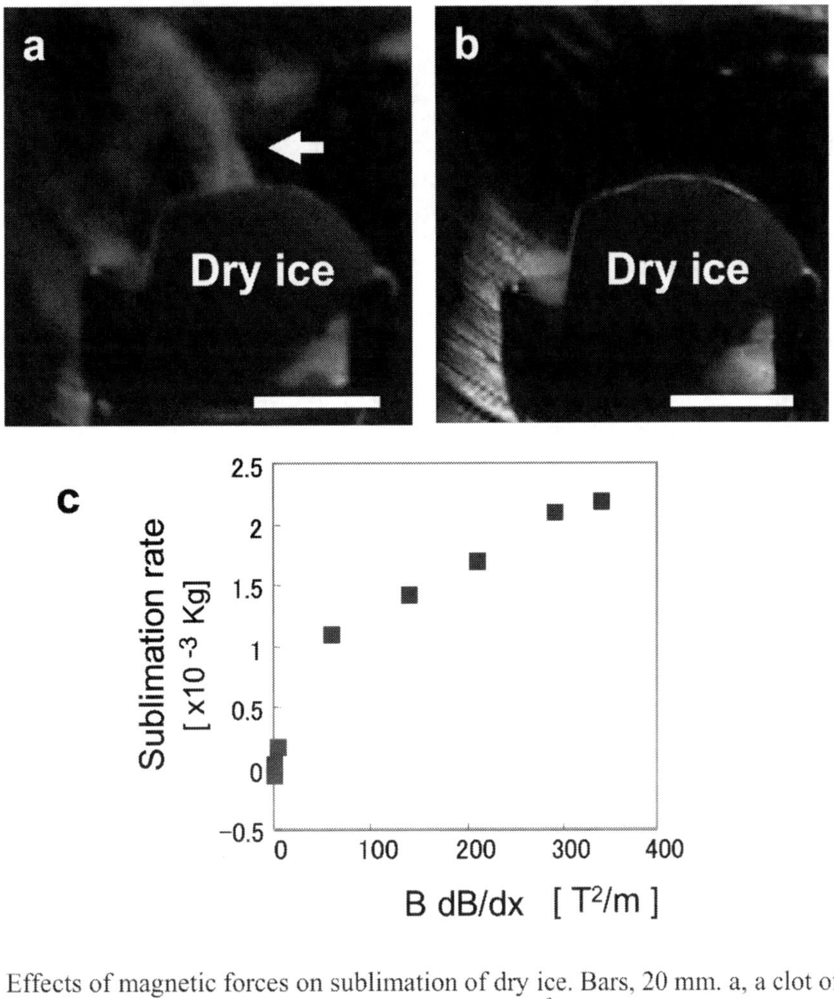

Figure 1. Effects of magnetic forces on sublimation of dry ice. Bars, 20 mm. a, a clot of dry ice set in 7 Tesla gradient magnetic fields. b, the dry ice set in 5 x 10^{-3} Tesla magnetic fields. c, sublimation of dry ice with an initial weight of 6.7 to 6.9 g at 25°C per 8 min. B dB/dx is the product of magnetic fields and gradients.

The obtained sublimation rate was plotted versus the product of magnetic fields and gradients (B dB/dx), which was proportional to the magnetic force acting on materials in the fields. The temperature inside bore was controlled by a water jacket with a thermally controlled water circulation.

3. Results and Discussion

Figure 1 shows photos dry ice with and without magnetic field exposure in a horizontal bore of 8 T superconducting magnet. In general, sublimation of dry ice produces streams of CO_2 aerosols directed downward because the density of the low-temperature CO_2 gas is heavier than that of air. A magnetic field of 7 Tesla, which had a spatial gradient of 40 T/m, remarkably enhanced the sublimation of dry ice (Fig. 1a) compared with dry ice in the magnet bore at 5 mT without superconducting currents (Fig. 1b). The magnetic field of 7 T caused the CO_2 aerosols to move upward (Fig. 1a) and the CO_2 aerosol flowed toward the higher magnetic fields of 8 Tesla. In case with the exposure, the aerosol formed a white and broad stream, which is marked by a white arrow (Fig. 1a). Dry ice in a 7 T gradient field exhibited the most rapid decrease in weight (Fig. 1c).

Next, the same kind of experiment was carried out in a vertical bore of 5 T magnet. In the observation of dry ice sublimating to CO_2 gas, gaseous stream was visualized by white aerosol containing water, oxygen, and nitrogen as well as the sublimated carbon dioxide, as shown in Figure 2 and 3. The surface of dry ice set in a room temperature atmosphere was partly covered by white powders. The white powder was mainly consisted of CO_2 because they increased in volume when additional CO_2 gas was flowed into the magnet's bore.

It was remarkable that the dry-ice-powder accumulated in the side faced toward higher fields (Fig. 2). Also the CO_2 aerosol stream directed to a higher magnetic field which was similar to the observation in a horizontal bore at 7 T (Fig. 1). The vertical type of magnet's bore generated two maximum gradient magnetic fields (GMF) along the axis of bore; the upper part and the lower part produced diamagnetic force in the upward and downward direction, respectively. The observed phenomena are paradoxical because CO_2 molecules have diamagnetism.

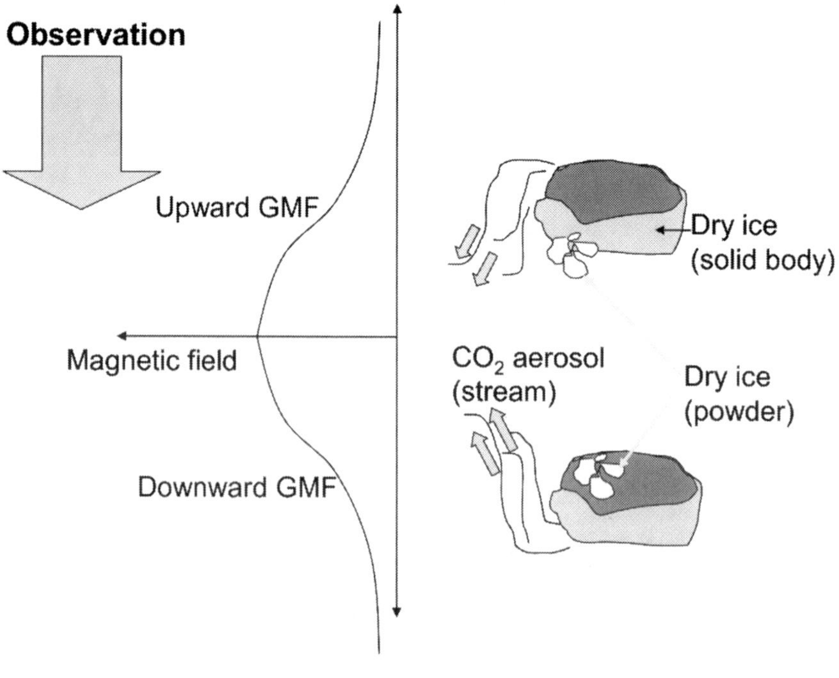

Figure 2. Behaviors of CO_2 aerosol and powder under gradient magnetic fields

Gas	He	O_2	CO_2	No additional gas (air only)
* Change in Sublimation	−	=	+	±0

* Qualitative evaluations (+, -) by eye

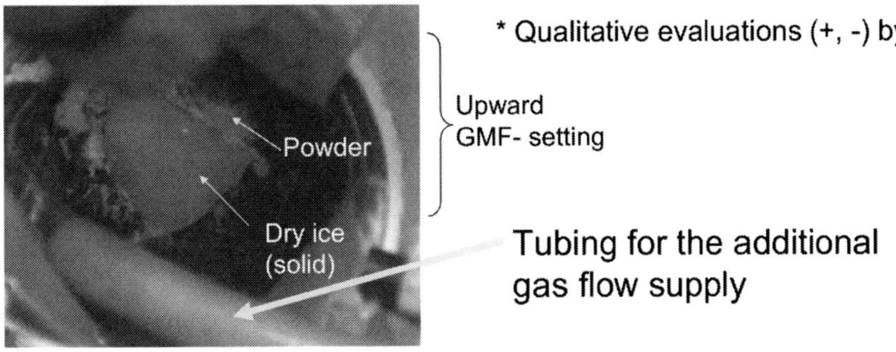

Figure 3. Effects of additional gas flow (He, O_2, and CO_2) under 4–5T gradient magnetic fields. The top panel (table) shows a qualitative evaluation for a relative change in dry ice sublimation by providing additional gas flows.

The magnetism of CO_2 is diamagnetism and a magnetic force directs CO_2 gas towards lower magnetic fields. However, oxygen gas, which is paramagnetic, has a magnetic force that is more than two 160 times greater than that of CO_2. Therefore, it is speculated that the oxygen gas carries the CO_2 aerosols towards the higher magnetic fields. In addition, the temperature gradient between the atmosphere surrounding the dry ice and areas at room temperature away from the dry ice provided a gradient in the magnetic susceptibility of the oxygen gases. The combination of magnetic field gradients and the susceptibility gradient enhance gas flow convection.

To investigate the possible effect of magnetic susceptibility of atmosphere surrounding the dry ice, behaviors of dry-ice-powder at the surface of fragment was observed when an additional gas, such as He, O_2, and CO_2, was supplied by tubing in a flow rate of ~2 ml/min shown in Fig. 3. The top panel (table) in Fig. 3 shows a relative change in dry ice sublimation rate during the additional gas supply. The frequency and amount of white aerosol stream flowing from the dry ice were evaluated by the experimenter's eyes qualitatively. The results showed the additional CO_2 supply enhanced the sublimation in dry ice; however, O_2 supply decelerated the sublimation. It was notable that no white powder appeared on the surface of dry ice during the additional O_2 supply. We can hypothesize that much of O_2 supply increased the magnetic susceptibility of atmosphere to a moderate condition where the oxygen transport in atmosphere became out of rate limiting.

In the future study, we need control the concentration of gases more preciously and clarify the mechanism for above mentioned difference between the additional gas supplies of CO_2 and O_2.

We can expect the mechanism may provide a new application of magnetic fields to the control of environmental CO_2 concentration by means of CO_2 collection by oxygen at the temperature of dry ice.

4. Conclusion

Enhancements of CO_2 sublimation in a fragment of dry ice, which was set in the bore of magnet, were observed. The data of sublimation rate suggest that the rate of dry ice sublimation depends on the magnetic field gradients. The sublimation enhancements were distinctly observed when the product of a magnetic field and its gradient, BdB/dx, was above 70 T^2/m.

Changing the gas concentration near the dry ice affected the magnetic field effect, and particularly supplying an additional gas of CO_2 and O_2 accelerated and decelerated the sublimation, respectively.

Acknowledgement

This work was partly supported by The Japan Securities Scholarship Foundation. The author wishes to thank Prof. S.Ueno for his great support.

References

[1] Melville D 1975 *Nature* **255** 706
[2] Friedlaender F J, Takayasu M, Rettig J B, Kentzer C P 1978 *IEEE Trans. Magn.* **14** 1158
[3] Ohara T, Mori S, Oda Y, Wada Y, Tsukamoto O 1995 *Journal of IEEJ* **116B** 979
[4] Fujiwara M, Kodoi D, Duan W, Tanimoto Y 2001 *J. Phys. Chem. B* **105** 3343
[5] Ikezoe Y, Hirota N, Nakagawa J, Kitazawa K 1998 *Nature* **393** 749
[6] Ikezoe Y, Kaihatsu T, Sakae S, Uetake H, Hirota N, Kitazawa K 2002 *Energy Conversion and Management* **43** 417
[7] Watarai H, Namba M 2002 *J. Chromatogr. A* **961** 3

Influence of strong magnetic field on the sedimentation of red blood cells

Y Tanimoto[1,2] **and Y Kakuda**[1]

[1]Graduate School of Science, Hiroshima University, Higashi-Hiroshima 739-8526, Japan
[2]Faculty of Pharmacy, Osaka Ohtani University, Tondabayashi 584-8540, Japan

E-mail: tanimoy@osaka-ohtani.ac.jp

Abstract. Influence of vertical magnetic fields of ±1200 T^2m^{-1} on the sedimentation of red blood cells were examined. The sedimentation rate changes 10-20 % by application of the magnetic fields. This result is interpreted in terms of the magnetic force acting on the cells.

1. Introduction

Recently it is known that magnetic field is a useful and convenient tool to manipulate chemical and physical processes [1]. Since 2002, our group has researched the influence of strong *vertical* magnetic field (15 T, 1500 T^2m^{-1}) on various chemical, physical and biological phenomena. As for physical phenomena, it was shown that X-ray crystallographic quality of *lysozyme* protein crystals is improved by crystallizing under magnetically simulated microgravity [2], photo-induced thermal convection of a benzene solution is controlled by application of magnetic field [3], plastic tips are separated using magnetically simulated microgravity [4], liquid and solid thin films, which cannot be prepared under earth's surface gravity, are prepared for the first time using magnetically simulated microgravity [5-7] and so on. It was also demonstrated that magnetic field can control three-dimensional morphological chirality in inorganic membrane tubes prepared by silicate garden reaction [8,9] and that magnetic field affects remarkably morphology of silver dendrite prepared by the solid-liquid redox reaction of copper metal and silver nitrate solution [10]. Furthermore, it was found that magnetic field influences dramatically the amphibian development [11].

Application of magnetic field to medical science is one of important topics in magneto-science. In 1987 Okazaki et al. reported that *paramagnetic* red blood cells moved to horizontal direction when *inhomogeneous horizontal* magnetic field (about 0.65 T, 40 T^2m^{-1}) was applied [12-15]. The results are interpreted in terms of magnetic force acting on paramagnetic cells, suggesting the magnetic separation of red blood cells. Influence of *homogeneous vertical* magnetic field (6.3 T) on sedimentation of red blood cells was reported [16,17]. The effect on the sedimentation rate of red blood cells in saline solution is explained in terms of magnetic orientation of the cells, whereas the effect in plasma is suggested due to the cell orientation and the magnetic-field-induced cell aggregation. The influence of simultaneous application of *homogeneous horizontal* magnetic field (8 T) and ac electric field (4 kV m^{-1}) on red blood cell sedimentation was attributed to the magnetic and electric orientation of the cells. [18]. Since then influence of magnetic fields on red blood cells has been studied extensively [19,20].

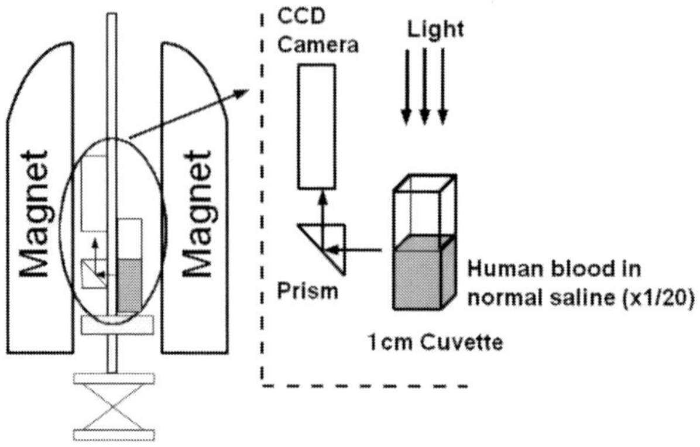

Figure 1. Experimental setup for sedimentation of red blood cells in magnetic fields.

In this paper, we have studied the influence of *inhomogeneous vertical* strong magnetic fields on the sedimentation of whole red blood cells in order to examine the possibility of their magnetic separation on a large scale. It is suggested that magnetic separation of red blood cells could be possible simply using a compact superconducting magnet.

2. Experimental

The experimental setup used in this study is schematically shown in Figure 1. A compact superconducting magnet (JASTEC, LH15T40; 15 T, 1500 T^2m^{-1} in a 40 mm vertical room temperature bore tube) was used [3]. Human blood, stored with anticlotting agent overnight, was diluted 20 times by normal saline (Otsuka Pharmaceutical) for optical observation. A quartz cuvette (10 mm × 10 mm × 40 mm) containing the 3-ml blood solution was placed in the room temperature bore tube. Magnetic field intensities were varied by changing the cuvette position in the bore tube. The interface between the red blood cells and the supernatant solution was observed by CCD camera from side [3], and the distance of the interface from the upper surface of the whole solution was analyzed by Scion Image.

3. Results and Discussion

Figure 2 shows movement of the interface between the red blood cell solution and the supernatant solution under magnetic fields. From the slope of the plot, the sedimentation rates are obtained as 6.5 × 10^{-2} (0 T^2m^{-1}, 15 T), 7.9 × 10^{-2} (+1200 T^2m^{-1}, 11 T) and 5.7 × 10^{-2} mm min^{-1} (-1200 T^2m^{-1}, 13 T), as shown in Table 1. Hereafter the product ($B\partial B/\partial z$) of the magnetic flux density (B) and its gradient ($\partial B/\partial z$) is called magnetic force field, the direction of z being vertical and upward. Comparing to the rate at 0 T^2m^{-1} and 15 T, the rate at increases about 20% by application of magnetic field of +1200 T^2m^{-1} and 11 T, whereas it decreases about 14% by application of that of -1200 T^2m^{-1} and 13 T.

Table 1. Sedimentation rate of red blood cells and calculated forces acting a red blood cell.

$B\partial B/\partial z$	B	Sedimentation rate	Force acting on a cell[a]
T^2m^{-1}	T	10^{-2} mm min^{-1}	10^{-10} N $cell^{-1}$
0	15	6.54	-1.01
1200	12	7.85	-1.23
-1200	12	5.65	-0.79

[a]Forces were calculated using eq. 1. See text.

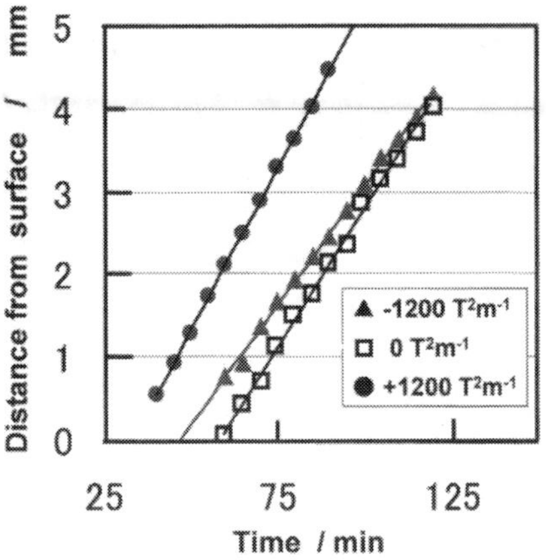

Figure 2. Plot of distance of the interface between red blood cell solution and supernatant solution from the upper surface of the whole solution *vs* time.

Magnetic orientation of red blood cells affects the sedimentation rate because of their disk-like shape, as already reported [16,17]. In this experiment, however, it is reasonable to think that the observed effect could be attributable to magnetic force acting on red blood cells, since all the experiment is carried out under high magnetic fields (11-15T) where magnetic orientation is saturated [21].

The force ΔF acting on a red blood cell in a magnetic force field is given by (1),

$$\Delta F = -(\rho_{RBC} - \rho_{NS})gV_{RBC} + (1/\mu_0)(\chi_{RBC} - \chi_{NS})V_{RBC}B\partial B/\partial z \quad (1)$$

where ρ_{NS} and ρ_{RBC} respectively denote the densities of normal saline and a red blood cell, respectively, g is the gravitational acceleration constant of the earth at surface, V_{RBC} is the volume of a red blood cell, χ_{NS} and χ_{RBC} are the respective volume magnetic susceptibilities of normal saline and a red blood cell and μ_0 is the magnetic permeability in vacuum. The first term of the right-hand side of eq. 1 is the force resulting from earth's surface gravity and the second term is the magnetic force attributable to the magnetic susceptibility difference between normal saline and the red blood cell.

The direction of the magnetic force changes depending on to the signs of two parameters (i.e., $B\partial B/\partial z$ and $(\chi_{RBC} - \chi_{NS})$). In the case of $B\partial B/\partial z > 0$, the magnetic force is downward when $\chi_{RBC} < \chi_{NS} < 0$, enhancing segmentation, while it is upward when $\chi_{NS} < \chi_{RBC}$, depressing sedimentation. In the case of $B\partial B/\partial z < 0$, the force is upward when $\chi_{RBC} < \chi_{NS} < 0$, while it is downward when $\chi_{NS} < \chi_{RBC}$. The experimental result, shown in Table 1, indicates $\chi_{RBC} < \chi_{NS} < 0$. Volume magnetic susceptibilities of an oxygenated red blood cell and a deoxygenated cell are reported as $-4\pi \times 0.736 \times 10^{-6}$ and $-4\pi \times 0.472 \times 10^{-6}$, respectively [22]. If we can assume that magnetic susceptibility of normal saline is the same as that of water ($-4\pi \times 0.715 \times 10^{-6}$ [23]), since the concentration of NaCl is as small as 0.9 wt%, then we can safely conclude that the red blood cells used here are mainly oxygenated ones (diamagnetic) as $\chi_{RBC} < \chi_{NS}$.

So the ΔF values are calculated using $\rho_{NS} = 1.00 \times 10^3$ kg m^{-3}, $\rho_{RBC} = 1.12 \times 10^3$ kg m^{-3} [24], $g = 9.80$ m s^{-2}, $V_{RBC} = 86.1 \mu m^3$ [24], $\chi_{NS} = -4\pi \times 0.715 \times 10^{-6}$, $\chi_{RBC} = -4\pi \times 0.736 \times 10^{-6}$ and $\mu_0 = 4\pi \times 10^{-7}$ $T^2J^{-1}m^3$. The results are shown in Table 1. At 0 T^2m^{-1}, a red blood cell settles out by the earth's surface gravity of -1.0×10^{-10} N cell^{-1}. In the presence of the magnetic force field, it settles out by the forces of -1.2×10^{-10} N cell^{-1} (+1200 T^2m^{-1}) and -0.80×10^{-10} N cell^{-1} (-1200 T^2m^{-1}). The agreement between observed segmentation rates and calculated forces in magnetic field is fairly good, as the observed sedimentation rate changes are 10-20% of the value at 0 T^2m^{-1}. Small difference between the upward and downward magnetic forces could be attributable to experimental error.

In conclusion, the sedimentation rate of red blood cells increases or decreases 10-20% by application of a magnetic force field of ±1200 T^2m^{-1}. The result is explainable semi-quantitatively by the magnetic force acting on the cells. The magnetic force of 1200 T^2m^{-1} to whole red blood cells corresponds 10-20% of the earth surface gravity, though the magnetic susceptibility difference between normal saline and red blood cell is as small as $4\pi \times 0.021 \times 10^{-6}$. Present result suggests that magnetic separation of red blood cells on a large scale is capable simply using a compact superconducting magnet, though further detailed study is necessary.

Acknowledgements
Authors thank Prof. Yoshihisa Fujiwara and Dr. Masao Fujiwara, Hiroshima University, for their valuable discussion.

References

[1] Yamaguchi M and Tanimoto Y (eds.) 2006 *Magneto-Science* (Tokyo: Kodansha/Springer)
[2] Yin D C, Wakayama N I, Harata K, Fujiwara M, Kiyoshi T, Wada H, Niimura N, Arai S, Huang W D, and Tanimoto Y 2004 *J. Cryst. Growth* **270** 184
[3] Koyama F and Tanimoto Y 2006 *Mol. Phys.* **104** 1703
[4] Tanimoto Y, Fujiwara M, Sueda M, Inoue K, and Akita M 2005 *Jpn. J. Appl. Phys.* **44** 6801
[5] Sueda M, Katsuki A, Nonomura M, Kobayashi R, and Tanimoto Y 2007 *J. Phys. Chem. C* **111** 14389
[6] Tanimoto Y, Yamamoto H, Fujiwara Y, Fujiwara M, and Katsuki A 2008 *Chem. Lett.* **37** 500
[7] Katsuki A, Yamamoto H, Fujiwara Y, Fujiwara M, Endo M, and Tanimoto Y 2008 *Chem. Lett.* **37** 728
[8] Uechi I, Katsuki A, Dunin-Barkovskiy L, and Tanimoto Y 2004 *J. Phys. Chem. B* **108** 2527
[9] Duan W, Kitamura S, Uechi I, Katsuki A, and Tanimoto Y 2005 *J. Phys. Chem. B* **109** 13445
[10] Tang F, Katsuki A, and Tanimoto Y 2006 *Mol. Phys.* **104** 1667
[11] Kawakami S, Kashiwagi K, Furuno N, Yamashita M, Kashiwagi A, and Tanimoto Y 2006 *Jpn. J. Appl. Phys.* **45** 6055
[12] Okazaki M, Maeda N, and Shiga T 1987 *Experientia* **42** 842
[13] Okazaki M and Shiga T 1987 *Eur. Biophys. J.* **14** 139
[14] Okazaki M, Kon K, Maeda N, and Shiga T 1988 *Physiol. Chem. Phys. Med. NMR* **20** 3
[15] Okazaki M, Seiyama A, Kon K, Maeda N, and Shiga T 1991 *J. Colloid Interface Sci.* **146** 590
[16] Iino M 1997 *Bioelectromagnetics* **18** 215
[17] Iino M and Okuda Y 2001 *Bioelectromagnetics* **22** 46
[18] Suda T and Ueno S 1999 *J. Mag. Soc. Jpn.* **23** 931
[19] Zborowski M, Ostera G R, Moore L R, Milliron S, Chalmers J J and Schechter A N 2003 *Biophys. J* **84** 2638
[20] Furlani E P 2007 *J. Phys. D: Appl. Phys.* **40** 1313 and references therein.
[21] Higashi T, Yamagishi A, Takeuchi T, Kawaguchi N, Sagawa S, Onishi S and Date M 1993 *Blood* **82** 1328
[22] Spees W M, Yablonskiy D A, Oswood M C and Ackerman J J H 2001 *Magn. Reson. Med.* **45** 533
[23] Chemical Society of Japan (ed.) 1966 *Kagaku-Binran*, (Tokyo: Maruzen) p.1080.
[24] Imabori K and Yamashita T (supervised) 1984 *Seikagaku-Jiten* (Tokyo: Tokyo-Kagakudojin) p.695

Study on Magnetic Separation System Using High Tc Superconducting Bulk Magnets for Water Purification Technique

T Oka[1], H Kanayama[1], K Tanaka[1], S Fukui[1], J Ogawa[1], T Sato[1], M Yamaguchi[2], M Ooizumi[1], K Yokoyama[3] and K Noto[4]

[1] Niigata University, 8050 Ikarashi-Nino-cho, Nishi-ku, Niigata 950-2181, Japan
[2] Japanese Super-conductivity Organization CO., LTD., 2-1-6 Etchujima, Koto, Tokyo, 135-8533 Japan
[3] Ashikaga Institute of Technology, 268-1 Ohmae-cho, Ashikaga, Tochigi, 326-8558 Japan
[4] Professor Emeritus Iwate University, 3-19-27 Chomeigaoka, Izumi-ku Sendai, 981-3212 Japan
E-mail: okat@eng.niigata-u.ac.jp

Abstract. The application of superconducting bulk magnets to the magnetic separation techniques has been investigated for the Mn-bearing waste water drained from the university laboratories. The research has been conducted in comparison with the electromagnets, and the cryo-cooled superconducting solenoid magnet. The separation ratios of ferrite precipitates including Mn element in the waste slurry were estimated by means of the high gradient magnetic separation method with ferromagnetic iron filters in the water channel and open gradient magnetic separation without them. As the magnetic force acting on the particles is given by the product of a magnetization of particles and a gradient of magnetic field, and a superconducting bulk magnet shows a sharp gradient of the magnetic field on the surface, the performances of the bulk magnet system were almost equivalent to those of the superconducting solenoid magnet with wide bore with respect to the magnetic separation ratios. The separation ratios for Mn have reached over 80 % for HGMS and 10 % for OGMS under the flow rates less than 3 liter/min.

1. Introduction

A water purification process has been operated on the waste water bearing toxic heavy metallic ions such as Mn, Zn, Cu, and Hg, which is drained from the university laboratories (3 tons/year). Figure 1 shows a conventional water purification process using the magnetic separation technique. The ferrite precipitates are formed by adding iron compounds to the waste water and subsequently removed by the magnetic field generated by permanent magnets. A rotating plate with hundreds of permanent magnets attracts the magnetic precipitates from the slurry water containing Fe ions, and the sludge is scraped away the vessel. Since a large amount of additives containing iron compounds have been used to throw them away, the reduction of additives, like iron sulphate and so on, in the process is necessary to reduce the cost to treat the waste water.

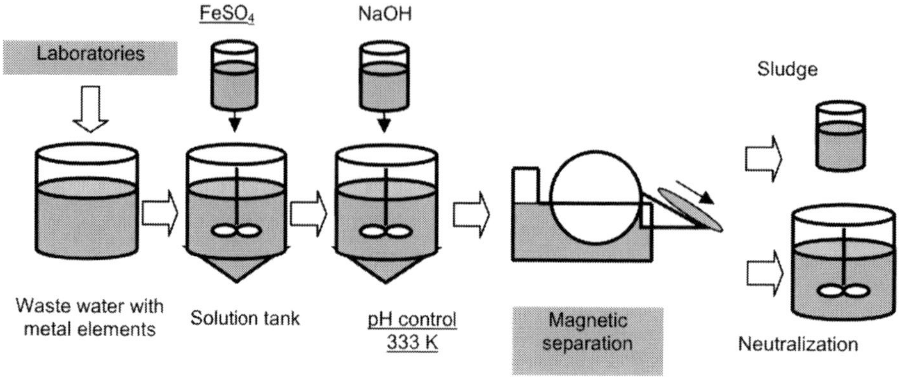

Figure 1. Conventional purification process for waste water including heavy metal elements

It has become well known that the melt-processed $REBa_2Cu_3O_y$ (RE123; Rare Earth = Y, Sm, Gd) high temperature superconductors (HTS) bearing RE_2BaCuO_5 (RE211) particles act as permanent magnets when they capture the magnetic fields [1, 2]. The field trapping ability of the superconducting bulk magnets has been substantially improved by synthesizing large grain textured bulk sample and reinforcing the sample to prevent against the fracture that happens due to the magnetic force when they trap the intense magnetic flux density of over several T. As was reported by M. Tomita et al. [3], the maximum trapped field attained in the Y123 bulk sample has reached 17.24 T at 29 K after an activation procedure in 18 T. On the other hand, various kinds of the superconducting permanent magnets using bulk superconductors have been constructed in order to develop the practical and industrial applications [4].

The magnetic force F_m is given by the product of three elements of volume, susceptibility and magnetic field, as is expressed in equation (1). We note that the presence of steep gradient is necessary to obtain high magnetic force in addition to the field strength [5].

$$\boldsymbol{F_m} = \frac{4}{3}\mu_0 \pi r_p^{\,3} \frac{9(\chi_p - \chi_f)}{(3+\chi_p)(3+\chi_f)} H \cdot \mathrm{grad}H \tag{1}$$

A couple of magnetic separation systems, high gradient (HGMS) and open gradient (OGMS) magnetic separation, have been precisely studied by Okada et al. [6] and Fukui et al. [7], as shown in

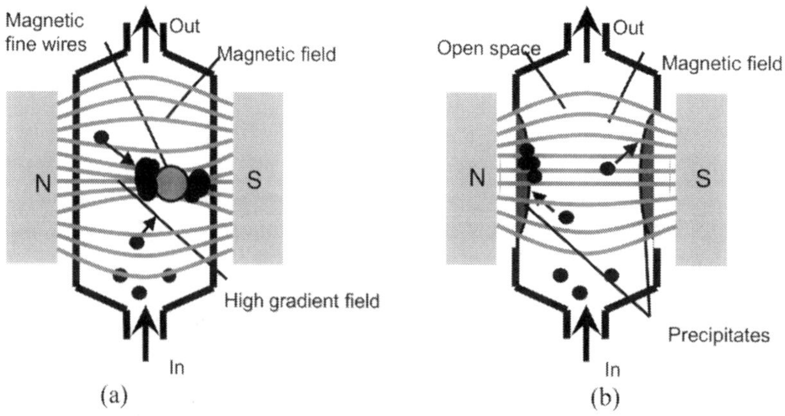

Figure 2. Magnetic separation techniques, (a) high gradient and (b) open gradient magnetic separation.

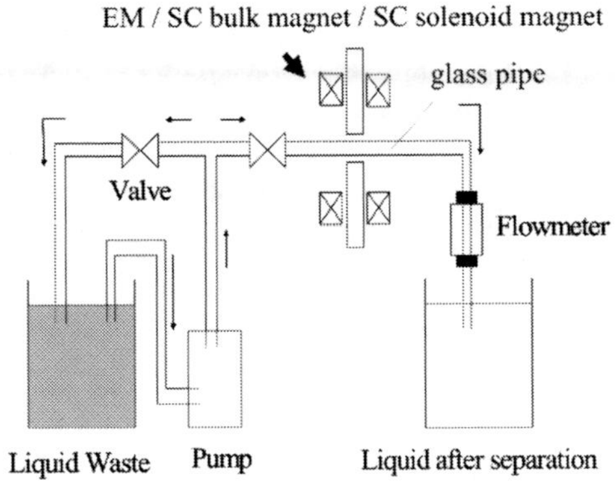

Figure 3. Experimental set up

Fig. 2. When ferromagnetic fine wires are inserted in the magnetic field, the flux lines between the magnetic poles are attracted to them, changing their distribution, resultantly yielding steep gradient of magnetic field around them (see Fig. 2 (a)). We call the system as HGMS. On the other hand, we know that the magnetic field generated between the poles originally possesses a steep gradient (Fig. 2 (b)). The later is ordinarily used in magnetic separation techniques called as OGMS.

According to the qualitative analysis by ICP, the waste water includes Fe, Mn, B, Ni, Zn, Cr, Cd, Mo, Co, and Cu ions. In this report, we focus on Fe and co-precipitated Mn contents that have major concentrations among them. And we aim to evaluate the separation performances on them by means of a superconducting bulk magnet, a superconducting solenoid magnet, and a conventional electromagnet.

2. Experimental

It is known that Mn ions co-precipitate with iron atoms when $FeSO_4$ is added. The precipitates were formed by adding $FeSO_4$ powder and NaOH (until pH 9.5), and then heating to 333-343 K. The mean precipitate size was 8.535 micrometer which was measured by Laser Diffraction Particle Size Analyzer SALD-3000J (Shimadzu Corp.).

A schematic figure of the experimental setup of magnetic separation is shown in Fig. 3, where SC and EM mean "superconducting" and "electromagnet", respectively. 5 liters of waste water including 91.6 ppm Fe and 5.85 ppm Mn was circulated and a part of it was led to the glass channel (20 mm in diameter) settled in the magnetic field, where the electromagnet and superconducting bulk magnet

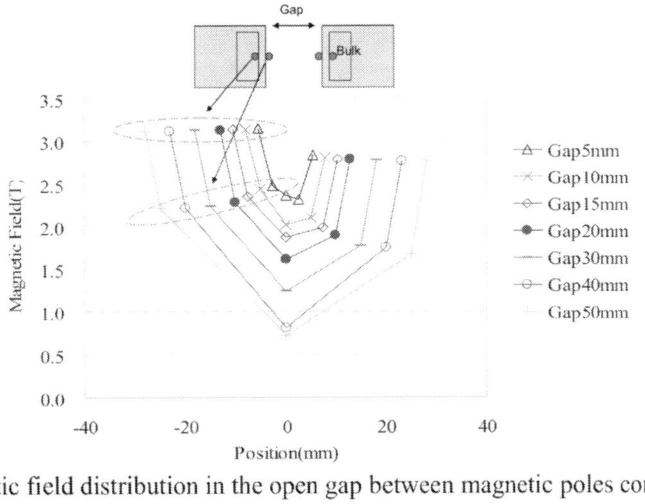

Figure 4. Magnetic field distribution in the open gap between magnetic poles constructed by face-to-face HTS bulk magnets.

generate up to 1.2 and 2.3 T at the surface of each pole surface. When we use a superconducting solenoid magnet, the channel was settled through the bore (100 mm in diameter), and 5 T magnetic field of was applied to it. The stainless steel mesh (SUS430 standardized in Japanese Industrial Standards) with a diameter of 0.1 mm was inserted into the channel with a volume density of 8.1 % for the HGMS, whereas the channel was settled without any filters in it for the OGMS. The separation ratio S % is defined as S=100(c_0-c_1)/c_0, where c_0 presents the initial ion concentration of whole waste water, c_1 shows that of clear water after the magnetic separation process.

The magnetic field distribution of a face-to-face type superconducting bulk magnet system is shown in Fig. 4, as a function of the gap between the magnetic poles, which consists of a pair of Sm-Ba-Cu-O bulk samples with a size of 60 mm in diameter, and they are mounted on the respective cold stages of the GM refrigerators in each vacuum vessel [8, 9]. The bulk samples were cooled to 35 K and magnetized by feeding pulsed currents by a condenser bank through the magnetizing pulse coils. The capacity of the power supply was 60 mF and the rise time of the magnetic pulses was 10 ms. The trapped field distribution was measured by scanning a Hall sensor (F. W. Bell BHA921) on the surface of each magnetic pole. As shown in Fig. 4, the performance of magnetic property has exhibited over 2 T at the surfaces of magnetic poles and has already exceeded the maximum values of the conventional permanent magnets such as Nd-Fe-B and even that of the large scale electromagnets. The magnetic field gradients are 173 T/m, 43.5 T/m, and 14.7 T/m for the bulk magnet, the superconducting solenoid, and the electromagnet, respectively.

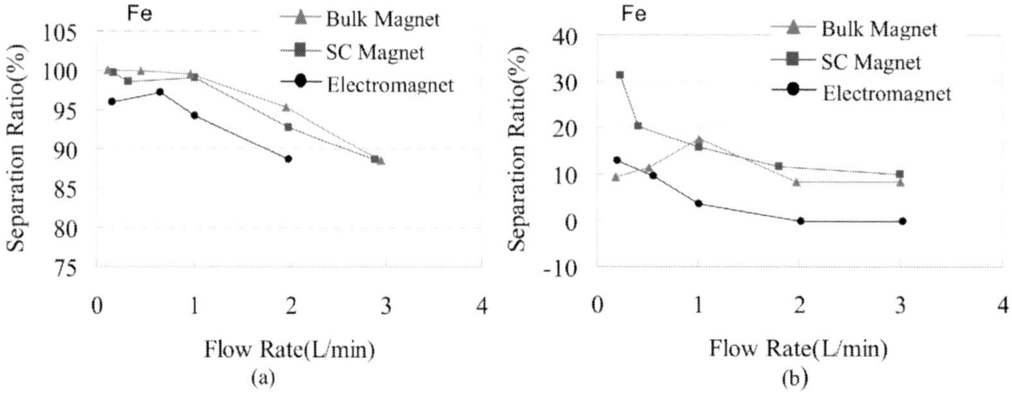

Figure 5. Separation ratios of Fe precipitates after (a) HGMS and (b) OGMS.

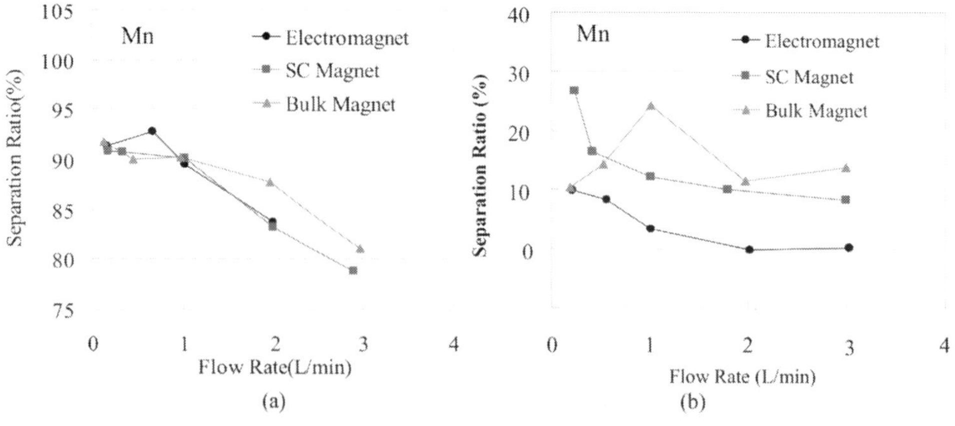

Figure 6. Separation ratios of Mn-bearing precipitates after (a) HGMS and (b) OGMS

3. Results and discussions

Figures 5 shows the separation ratios of Fe, large amount of which is added to co-precipitate heavy ions such as Mn, Zn, Cu and Hg in the disposal process. The analyzed data are plotted against the flow rates of slurry water with respect to the HGMS (a) and OGMS (b), respectively. It may be easy to understand that the separation ratios of HGMS exhibit the data more than 85 % whereas that of OGMS showing less than 30 % and they decrease with increasing flowing rate up to 3 liter/min. The data of bulk magnet and SC magnet occupy the superior region to that of electromagnet. It must be noted that the lines come close to each other without showing any substantial differences between the bulk magnet and SC magnet. This implies that the steeper magnetic field gradient residing around the bulk magnet attributes to this phenomena in spite of the narrower space than that of SC magnet.

Figure 6 shows the separation ratios of Mn, which co-precipitates with Fe, against the flow rates with respect to three magnetic field generators in the case of HGMS (a) and OGMS (b), respectively. The disposal processes should be originally conducted with a purpose to remove this heavy element from the waste water. As the data generally follow the tendency of Fe shown in Fig. 5, the separation ratios drastically decreases with increasing flow rates of the water. The fact that the performances of separation are slightly inferior to those of Fe can be easily explained by the reason that the Mn ions are not perfectly involved to the Fe-based precipitates. Since the stronger magnetic field would have an advantage in attracting the ferromagnetic precipitates, performances of bulk magnet and SC magnet should be superior to that of electromagnet. On the other, it may be expected that the data among three magnetic field generators would have no substantial differences in the regions where the flow rate is relatively low. The superconducting bulk magnets are characterized as compact and strong magnets with steep gradient, whereas the superconducting solenoid magnets are featured as field generators with vast spaces and uniform fields. We must choose the most appropriate device to adapt to the process depending on the practical conditions such as size, cost and so on.

4. Conclusions

The ferromagnetic precipitates containing Mn in the waste water drained from the laboratories were effectively removed by magnetic separation techniques with a trapping efficiency over 80 % for HGMS and 10 % for OGMS experiment using superconducting bulk magnets, which was conducted under the region of flow rates up to 3 liter/min. As the performances of superconducting bulk magnet and superconducting solenoid magnet are estimated as approximately equal in spite of the size of exposure space and time, the superconducting bulk magnets can be adapted to the disposal processes for Mn ions as well as large scale superconducting solenoid magnet. This attributes to the steep magnetic gradient of a bulk magnet system, which apparently is an advantage to other strong magnetic field generators.

References

[1] Weinstein R, Chen In-Gann, Liu J and K Lau 1991 *J. Appl. Phys.* **70** 6501
[2] Krabbes G, Fuchs G, Schatzle P, Gruss S, Park J, Hardinghaus F 2000 *Advances in Superconductivity XII Tokyo Springer-Verlag* 437
[3] Tomita M and Murakami M 2003 *Nature* **421** 517
[4] Oka T, Yokoyama K and Noto K 2005 *Supercond. Sci. Technol.* **18** S72
[5] Watson J 1973 *J. Appl. Phys.* **44** 4209
[6] Okada H, Kudo Y, Nakazawa H, Chiba A, Mitsuhashi K, Ohara T and Wada H 2004 *IEEE Trans. Appl. Supercond.* **14** 1576
[7] Fukui S, Takahashi Y, Yamaguchi M, Sato T, Imaizumi H, Ooizumi M, Nishijima S Watanabe T 2004 *IEEE trans. Appl. Supercond.* **14** 1568
[8] Oka T, Itoh Y, Yanagi Y, Yoshikawa M, Ikuta H Mizutani U 2000 *Physica C* **335** 101
[9] Oka T, Yokoyama K, Okada H, Katagiri K, Noto K, Itoh Y, Ikuta H and Mizutani U 2003 *IEEE Trans. Appl. Supercond.* **13** 1584

Study on magnetic separation of nanosized ferromagnetic particles

R Nakao[1], Y Matuo[1], F Mishima[1], T Taguchi[2,3] and S Nishijima[1]

[1] Division of Sustainable Energy and Environmental Engineering, Graduate School of Engineering, Osaka University, Suita, Osaka 565-0871, Japan
[2] Department of Biochemistry, Osaka University Medical School, Suita, Osaka, 565-0871, Japan
[3] Department of Cell Biology, Institute for Molecular Bioscience, University of Queensland, 4072, QLD, Australia

E-mail: ryo-nakao@qb.see.eng.osaka-u.ac.jp

Abstract. In recent researches in medicine and the pharmaceutical sciences, the magnetic separation technology using nanosized ferromagnetic particle is essential. For example, in the field of cell engineering, magnetic separation of nanosized ferromagnetic particles is necessary, but separation technology of nanosized particle using magnetic force has not been established. One of the reasons is that magnetic force acting on the object particles decreases as particle diameter becomes small, and makes magnetic separation difficult. In this study, magnetic force acting on the separation object was enlarged by the combination of superconducting magnet and the filter which consists of ferromagnetic particle. As a result of particle trajectory calculation and magnetic separation experiment, it was confirmed that the ferromagnetic particles of 15nm in diameter can be trapped in the magnetic filter under an external magnetic field of 0.5T. The ferromagnetic particles of 6nm in diameter which could not be separated under the same condition could also be trapped under 2.0T of external magnetic field.

1. Introduction

In recent researches in medicine and the pharmaceutical sciences, magnetic separation technology using nanosized ferromagnetic particle has gained importance [1]. In the field of cell engineering, nanosized ferromagnetic particles of several nanometers to 20nm in diameter is required, and magnetic separation technology of nanosized particle using magnetic force must be established [2-5]. However, magnetic force acting on the object ferromagnetic particle decreases as particle diameter becomes small, and makes magnetic separation difficult [6].

In this study, to enlarge magnetic force acting on the object ferromagnetic particles, we tried to design a magnetic filter consisting of densely packed ferromagnetic particles of 0.3mm in diameter in the glass tube of 4mm in inner diameter. First, two dimensional finite element model of the magnetic filter was constructed, and the magnetic field analysis and the fluid analysis were performed. Based on these results, the particle trajectory of the ferromagnetic particles of ϕ=6nm and ϕ=15nm in the vicinity of the magnetic filter were calculated by solving the dynamic equation of the object ferromagnetic particle in order to examine the possibility of magnetic separation. Next, to verify the result of simulation, magnetic separation experiment of ferromagnetic particles of ϕ=6nm (FePt) [7] and ϕ=15nm (Fe_3O_4) using the developed filter was conducted, and possibility of separation in the actual system was studied.

2. Experiment

2.1. The magnetic field analysis and the fluid analysis of the magnetic filter

The distribution of magnetic field intensity and flow velocity distribution in the model calculated by using ANSYS®Ver.10.0(ANSYS. Inc.) is shown in Figure 1. As shown in Figure 2, two dimensional finite element model was constructed based on maximum distance between two particles in three dimensional close-packed structure of spherical particle. Experimental conditions are shown below.

The maximum magnetic flux density at the center of coil were set to 0.5T and 2.0T, the particle diameter of the object ferromagnetic particles was set to 6nm and 15nm, the fluid was set to water (viscosity=1cP), and inflow velocity was set to 0.01m/s. The filter consisted of uniformly arranged ferromagnetic particles of 0.3mm in diameter (Figure1).

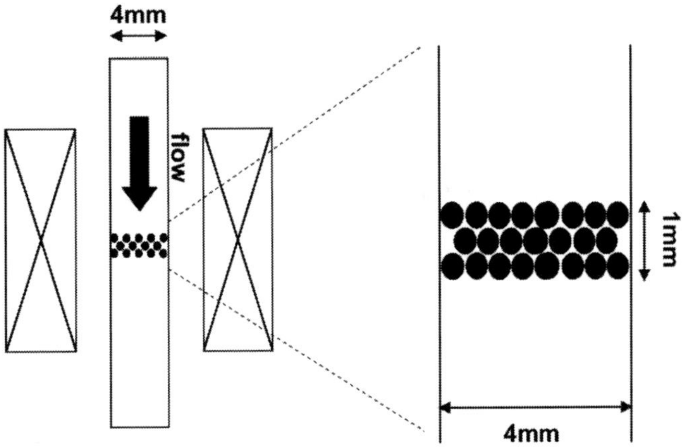

Figure 1. A schematic view of magnetic separation system.

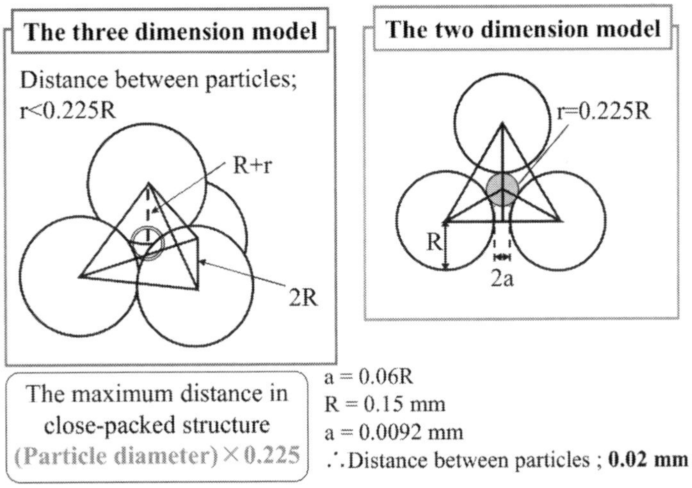

Figure 2. The method of making two dimension model.

2.2. Calculation of the particle trajectory of the ferromagnetic particles in the vicinity of magnetic filter

Magnetic force and drag force acting on the object particles were calculated based on the parameters obtained from the numerical analysis in each node using the magnetic field analysis and the fluid analysis using ANSYS. Motion equations of the object particles in the node using those forces were solved, the acceleration was calculated, and the next node was searched. The particle trajectory of the ferromagnetic particles in the vicinity of magnetic filter was calculated by repeating this operation.

2.3. Synthetic procedure of nanoparticles

FePt and Fe_3O_4 were supplied by Hitachi Maxell, Ltd.. The FePt nanoparticles were synthesized by reduction of platinum acetylacetonate and decomposition of iron pentacarbonyl in the presence of oleic acid and oleyl amine. The Fe_3O_4 nanoparticles were synthesized by the pyrolysis of iron carboxylate salts.

2.4. Magnetic separation experiment of nanosized ferromagnetic particles

To verify the result of simulation under 0.5T and 2.0T of external magnetic field in the experiment 2.1 and 2.2, magnetic separation experiment with actual the ferromagnetic particle was conducted. The developped magnetic filter was put under 0.5T and 2.0T of external magnetic field. Magnetic separation was performed by passing the fluid containing FePt (ϕ=6nm), Fe_3O_4 (ϕ=15nm) (Table 1) through the magnetic filter. The ferromagnetic particles were dispersed in n-hexane or toluene because nanosized particles are difficult to disperse uniformly in water. To measure the density of the ferromagnetic particles dispersed in organic solvent before and after the separation, the ferromagnetic particles were dissolved in 50% HCl after evaporating organic solvent, and measured by inductively-coupled plasma atomic emission spectrometry. The separation efficiency was estimated from change in the density of ferromagnetic particles.

Table 1. Properties of particle suspensions.

Materials	Particle diameter(nm)	Saturation magnetization(T)	Fluid(Viscosity)	Density(mg/ml)
Fe_3O_4	15	0.408	toluene(0.6 mPa·s)	0.1
FePt	6	1.144	n-hexane(0.3 mPa·s)	0.15

3. Results and discussion

3.1. The magnetic field analysis and the fluid analysis of the magnetic filter

As a result of the magnetic field analysis, the magnetic flux density and the magnetic gradient are higher in oblique direction between particles compared with horizontal direction between particles under both 0.5T and 2.0T of external magnetic field. Moreover, the magnetic flux density increased towards outside of the glass tube (Figure 3a). From the result of the fluid analysis, fluid speed near the top of the particles is slowest, and that at the center between the particles in horizontal direction is fastest (Figure 3b).

These results predict that the accumulation rate of the object particles on the magnetic filter is the highest in the vicinity of obliquely upward area of the magnetic filter towards the direction of the flow.

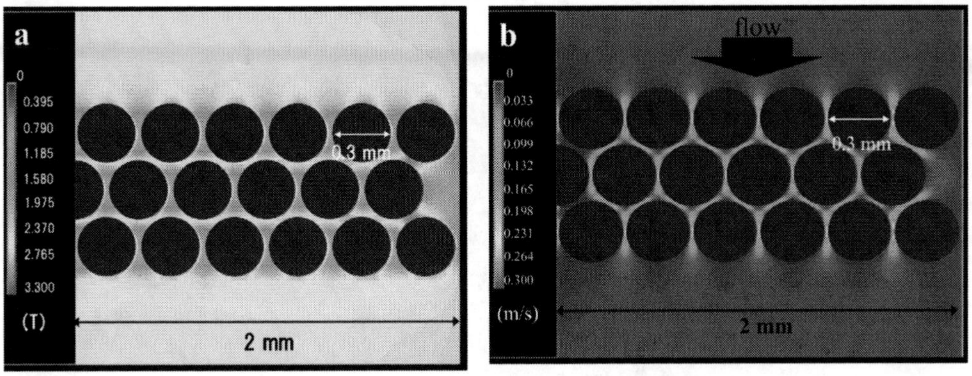

Figure 3a. The result of the magnetic field analysis (external magnetic field=2.0T).
Figure 3b. The result of the fluid analysis.

3.2. Calculation of the particle trajectory of the ferromagnetic particles in the vicinity of magnetic filter

The particle trajectory of the ferromagnetic particles of ϕ=6nm and ϕ=15nm showed that the ferromagnetic particles of ϕ=15nm could be trapped in the magnetic filter under 0.5T of external magnetic field (Figure 4a). In contrast, the ferromagnetic particles of ϕ=6nm could not be trapped in the magnetic filter under 0.5T of external magnetic field (Figure 4b). However, when external magnetic field was set to 2.0T, the ferromagnetic particles of ϕ=6nm also could also be trapped in the magnetic filter (Figure 4c).

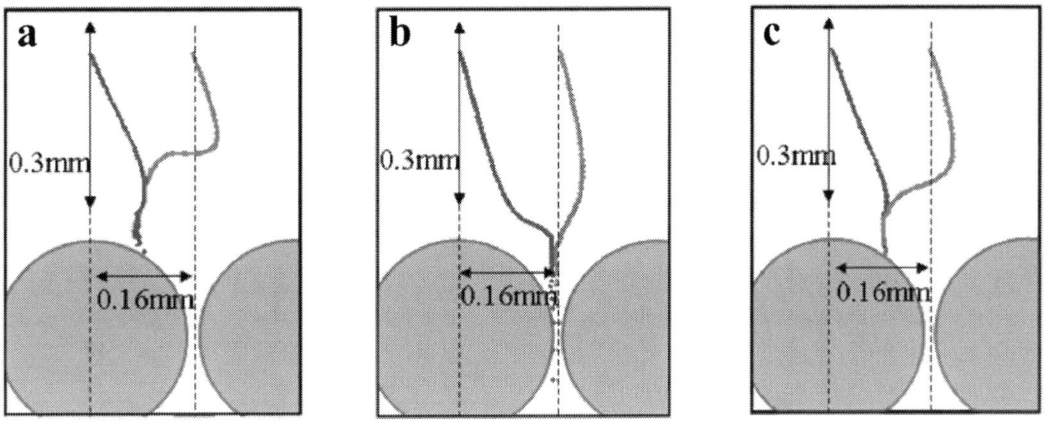

Figure 4. Calculation of the particle trajectory of the ferromagnetic particles.
(**a.** ϕ=15nm, external magnetic field=0.5T, **b.** ϕ=6nm, external magnetic field=0.5T,
c. ϕ=6nm, external magnetic field=2.0T)

3.3. Magnetic separation experiment of nanosized ferromagnetic particles

To verify the result of simulation discussed above, magnetic separation experiment under 0.5T and 2.0T of external magnetic field using FePt (ϕ=6nm) and Fe_3O_4 (ϕ=15nm) which are dispersed in n-hexane or toluene was conducted. Fe_3O_4 (ϕ=15nm) could be trapped in the magnetic filter under 0.5T of external magnetic field (Figure 5a). On the other hand, FePt (ϕ=6nm) could not be trapped in the magnetic filter under 0.5T of external magnetic field (Figure 5b), but it could be trapped in the magnetic filter under 2.0T of external magnetic field (Figure 5c).

Separation efficiency 94.3% Separation efficiency 3.7% Separation efficiency 40.0%

Figure 5. The Picture before and after magnetic separation experiment.
(**a.** ϕ=15nm, external magnetic field=0.5T、 **b.** ϕ=6nm, external magnetic field=0.5T、 **c.** ϕ=6nm, external magnetic field=2.0T)

4. Conclusion

The possibility of the magnetic separation using nanosized ferromagnetic particles was considered by calculation of the particle trajectory of nanosized ferromagnetic particles examined and the magnetic separation experiments of ferromagnetic particles.

As a result of particle trajectory calculation and magnetic separation experiment, it was confirmed that the ferromagnetic particles (ϕ=15nm) can be trapped in the magnetic filter consisting of densely packed ferromagnetic particles of 0.3mm in diameter in the glass tube of 4mm in inner diameter under 0.5T of external magnetic field. The ferromagnetic particles (ϕ=6nm) which could not be separated under the same condition could also be trapped under 2.0T of external magnetic field.

In this study, it was confirmed that the ferromagnetic particles which could not be trapped under external magnetic field using permanent magnet become possible to separate by increasing external magnetic field using superconducting magnet.

References

[1] Li H S, Stolz D B, Romero G 2005 *Traffic* **6** 324
[2] Lee J H, Huh Y M, Jun Y W, Seo J W, Jang J T, Song H T, Kim S, Cho E J, Yoon H G, Suh J S, Cheon J W 2007 *Nature Medicine* **13** 95
[3] Song H T, Choi J S, Huh Y M, Kim S J, Jun Y W, Suh J S, Cheon J W 2005 *J. Am. Chem. Soc.* **127** 9992
[4] Li J, Zhao X, Shi Y, Cai Y, Mou S, Jiang G 2008 *Journal of Chromatography* A **1180** 24
[5] Huh Y M, Jun Y W, Song H T, Kim S J, Choi J S, Lee J H, Yoon S, Kim K S, Shin J S, Suh J S, Cheon J W 2005 *J. Am. Chem. Soc.* **127** 12387
[6] Yavuz C T, Mayo J T, Yu W W, Prakash A, Falkner J C, Yean S, Cong L, Shipley H J, Kan A, Tomson M, Natelson D, Colvin V L 2006 *Science* **314** 964
[7] Sun S, Murray C B, Weller D, Folks L 2000 *Science* **287** 1989

Journal of Physics: Conference Series **156** (2009) 012033 doi:10.1088/1742-6596/156/1/012033

Removal and recycle of phosphate from treated water of sewage plants with zirconium ferrite adsorbent by high gradient magnetic separation

D Ito, K Nishimura and O Miura

Graduate School of Science and Engineering, Tokyo Metropolitan University,Hachioji, Tokyo, 192-0397, Japan

E-mail: dai@eei.metro-u.ac.jp

Abstract. Zirconium ferrite particles are good adsorbent for phosphate ions. Magnetic separation characteristics for removal of phosphate from treated water of sewage plants with the adsorbent have been studied to prevent eutrophication of semi-enclosed bay, e.g. the bay of Tokyo. Based on the adsorption for the phosphate ions and ferromagnetic properties of the zirconium ferrite adsorbent, high gradient magnetic separation characteristics with using superconducting magnet was discussed. Very rapid magnetic filtration velocity, i.e. 1m/s, and regeneration properties of the adsorbent indicate that the zirconium ferrite is the excellent adsorbent for phosphorus removal and recycle from treated water of large scale sewage plants.

1. Introduction

In the case of the bay of Tokyo, 5-6 tons of phosphate flow into the bay in the treated discharge water from the sewage disposal plants per day in spite of reducing phosphate to below the quality standard, i.e. 1 mg/L, by the advanced wastewater treatment methods A_2O. Fishes and shells so called "Edomae" are decreasing due to the red tide and the blue tide generated by eutrophication by nutrient inputs, i.e. phosphorus and nitrogen. Discharge water from sewage treatment plants is a major source of nutrient inputs into aquatic systems. Scientists of oceanograhy have proposed to further decreasing of phosphate to 60% of 1 mg/L to prevent the eutrophication. To meet their demands, we studied applicability of the high gradient magnetic separation HGMS technology with zirconium ferrite adsorbent to remove the posphate from the discharge water to less than 0.1 mg/L level. The HGMS with superconducting magnet not only has great potentiality to reduce the phosphate from the discharge water rapidly without generation of secondary products or sludge etc, i.e. zero-emission process, but possibility to reuse of trapped adsorbent.

Zirconium ferrite particles are good adsorbent for phosphate ions. National Institute for Environmental Studies NIES demonstrated the ability of the zirconium ferrite as the adsorbent of the phosphate in the Kasumigaura project and developed a phosphate recover station. Recycle of the phosphate is possible by washing zirconium ferrite particles with NaOH solution. The phosphate is , therefore, able to recycle, by the same method in the case of HGMS. USA prohibit export of phosphorous ore in 1997 and China is also going to stop export. Japan has no phosphorous resources, therefore, recycle of the phosphate in the discharge water from the sewage disposal plants is a very important challenge for us.

2. Experimental

Zirconium ferrite **$ZrFe_2(OH)_8$** adsorbent was offered by Japan EnviroChemicals Ltd. The adsorbent has sphere structure with 0.7 mm diameter for the filtration method, i.e. developed for the Kasumigaura project. The adsorbent particles contain not only zirconium ferrite particles with around 10 micron diameter but also binder to form the spherical shape.

SQUID magnetization measurement system MPMS-3 was used for magnetization measurements.

Experimental apparatus for the HGMS with a 10 T superconducting magnet having 10 cm room temperature bore diameter is shown in Fig.1. At the bore center, filter matrix of ferromagnetic stainless wire with 100 micron diameter was located.

Adsorption time dependence of the removal ratio for the phosphate in sample water to the adsorbent was measured.

Phosphate density in sample water was measured with a spectro photometer.

Typical magnetic separation experiment was conducted for discharge water offered from a sewage treatment plant under the condition with using 800 ppm zirconium ferrite adsorbent, 15 minutes adsorption time, at 10 T and at 0.03 L /sec flow rate. The phosphate density in the discharge water was 0.86 mg / L.

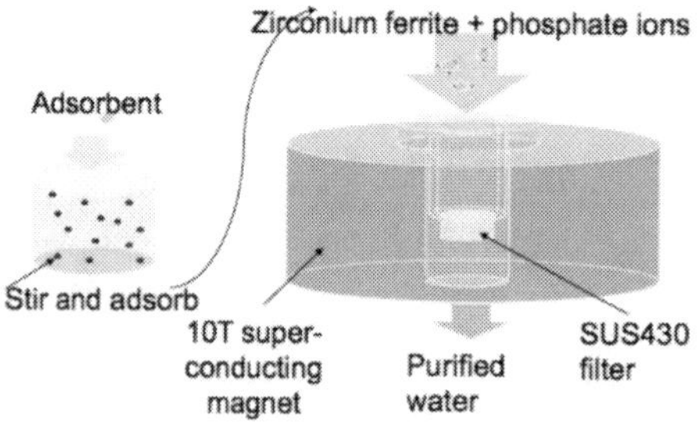

Figure 1. Experimental apparatus for HGMS.

3. Results and discussion

Zirconium ferrite adsorbent has ferromagnetic characteristics as shown in Fig.2. owing to Fe ions in the ferrite structure. Upper limit of the treatment water velocity for the HGMS, i.e. a magnetic velocity v_m, is given by

$$v_m = (2/9)(\mu_0 \chi M_s H_0 R^2 / a\eta) \quad (1)$$

where M_s is the saturation magnetization of the filter matrix, χ is the magnetic susceptibility of the adsorbent, η is the water viscosity, R is the adsorbent particle radius, a is matrix wire radius, H_0 is applied magnetic field[1]. In the case of ferromagnetic adsorbent, the saturation magnetization of the adsorbent M_a can be used instead of the χH_0. Rapid theoretical velocity of v_m =1 m/s is obtained by substitution M_a for χH_0.

Adsorption characteristic of the phosphate to the zirconium ferrite adsorbent is shown in Fig.3. The results indicate that more than 90% of phosphate in the discharge water can be removed in 5 minutes.

The HGMS could reduce the phosphate density in the discharge water offered by a sewage treatment plant from 0.86 mg / L to 0.006 mg / L under the typical condition. The removal ratio for the phosphate by the zirconium ferrite adsorbent is 99%. This result shows that the HGMS technology with the zirconium ferrite adsorbent is useful for preventing the eutrophication of semi-enclosed bay by it's introduction into sewage treatment plants. The technology can eliminate the advanced wastewater treatment A_2O process in a sewage treatment plant for reduction of phosphate.

Captured adsorbent to the filter matrix is easily detrapped from the matrix by decreasing magnetic field of a supronducting magnet. Phosphate ions adsorbed by zirconium ferrite particles can be removed with NaOH solution and the zirconium ferrite adsorbent is able to use repeatedly. Zero-emission process, therefore, can be achieved.

The principle of the NIES phosphate recover station developed for the Kasumigaura project is available for the recycle system of the phosphate.

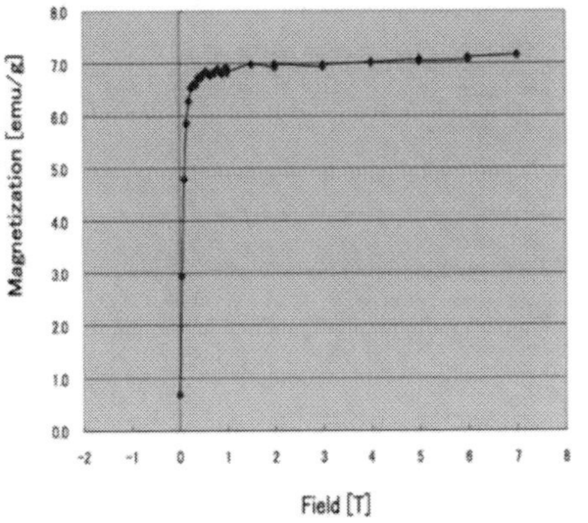

Figure 2. Magnetization of zirconium ferrite adsorbent.

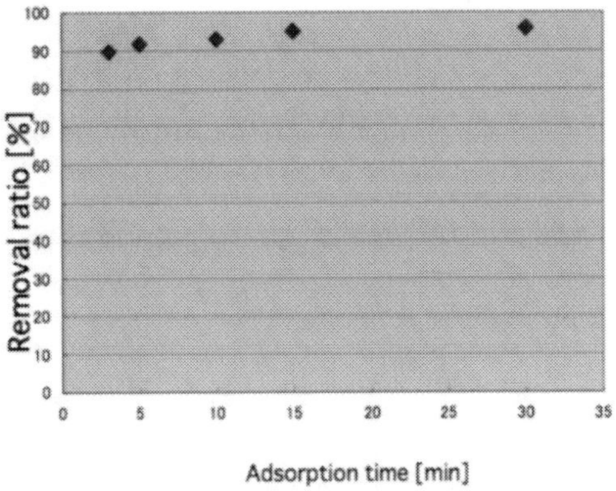

Figure 3. Adsorption characteristics of zirconium ferrite adsorbent.

4. Conclusion

Ferromagnetic properties revealed in this study show that the zirconium ferrite is excellent adsorbent of phosphate for HGMS.

High separation velocity of HGMS with superconducting magnet estimated from the magnetic velocity is suitable to remove phosphate from discharge water of large sewage treatment plants.

Zirconium ferrite is reusable adsorbent, therefore, phosphate removal in zero-emission is possible.

Phosphorus recycle is possible with using the principle of the NIES recover station.

We thank Japan EnviroChemicals Ltd. for giving us zirconium ferrite adsorbent SEVENTOOR-P.

References

[1] Watson J H P *et al.* 1996 *Minerals Engineering* **9** 973.

Magnetic separation of antibiotics by electrochemical magnetic seeding

I Ihara[1], K Toyoda[1], N Beneragama [2] and K Umetsu[2]

[1] Department of Agricultural Engineering and Socio Economics, Kobe University, Nada, Kobe 657-8501, Japan

[2] Department of Animal Science, Obihiro University of Agriculture and Veterinary Medicine, Obihiro, Hokkaido 080-8555, Japan

E-mail: ihara@port.kobe-u.ac.jp

Abstract. Magnetic separation of several classes of antibiotics was investigated using electrochemical magnetic seeding. Electrocoagulation with a sacrificial anode followed by addition of magnetite particles was applied for the magnetic seeding of antibiotics. With electrochemical magnetic seeding using an iron anode, tetracycline antibiotics (oxytetracycline, chlortetracycline, doxycycline and tetracycline) and cephalosporin antibiotic (cefdinir) were rapidly removed from synthetic wastewater by magnetic separation using a neodymium magnet. Iron and aluminium anodes were suitable for magnetic seeding of the antibiotics. The results indicated that the ability of antibiotics to form strong complex with iron and aluminium allowed the higher removal by magnetic separation. This method would be appropriate for rapid treatment of antibiotics in wastewater.

1. Introduction

Antibiotics are widely used to treat animal diseases and improve productivity in livestock farming. Administration of these may be performed as orally feed additives or directly by injections. The use of antibiotics may result in drug residues being present in animal products as well as in animal wastes. These animal wastes after excretion are carried into the environment through direct deposition, discharge from the treatment facilities, overflow or leakage from the storage facilities, or via direct land application through which the antibiotic residues as well as antibiotic resistance genes, are distributed into the environment simultaneously. This may be responsible for the promotion of resistant strains of bacteria in the environment [1]. Moreover, prolong human exposure to trace amount of antibiotic residues could cause chronic illnesses, especially in the hypersensitive individuals. In Japan, annual usage of antibiotics for animal medication is higher than that of human medication. The most widely used group of antibiotics in Japan's animal husbandry is tetracycline antibiotics. In recent years, environmental contaminants of antibiotics derived from livestock wastes are reported [2, 3]. Inappropriate use of medications has raised concerns about increased antibiotic resistance [4]. The presence of antibiotics and resistant bacteria are emerging environmental problems. Therefore, it is necessary to develop a certain method for treatment of antibiotics in animal wastes so that the safety of the environment can be assured.

Magnetic separation is a method, which uses the magnetic properties of the particles to be separated by magnetic force. The concept of using magnetic separation in wastewater treatment has been encouraged after the introduction of high-gradient magnetic separation (HGMS). The removal of non-ferromagnetic pollutants by magnetic separation is of greater importance for the environmental remediation. However, it is a well-known fact that magnetic separation has been used for the removal of ferromagnetic materials in wastewater. In this regard, it is required to enhance the magnetic susceptibility of the pollutants to be removed. Electrochemical magnetic seeding, which is performed using sacrificial anodes and magnetite particles to enhance the magnetic properties of non-ferromagnetic pollutants had been investigated [5]. For magnetic separation of antibiotics presented in animal wastes, it requires an efficient method of magnetic seeding. The objective of this study was to evaluate electrochemical magnetic seeding method for magnetic separation of antibiotics in aquatic solution. In order to enhance the magnetic force acting on antibiotics, the magnetic seeding was operated with electrocoagulation using iron electrodes followed by addition of magnetite powder.

2. Materials and methods

2.1. *Reagents*

Figure 1 shows chemical structures of antibiotics investigated in this study. Oxytetracycline hydrochloride, doxycycline hydrochloride, and levofloxacin hydrochloride were purchased from Wako Pure Industries (Osaka). Tetracycline hydrochloride, chlortetracycline hydrochloride and cefazoline sodium salt were purchased from Sigma-Aldrich Chemical. Cefdinir was purchased from Astellas Pharma Inc. (Osaka). Magnetite (Fe_3O_4) powder was obtained from Kanto Chemical Co. Inc (Tokyo).

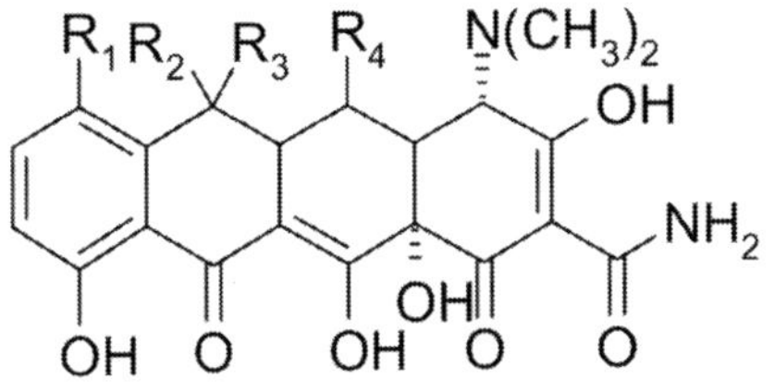

A: tetracycline antibiotics

Name	Symbol	R_1	R_2	R_3	R_4
tetracycline	TC	H	CH_3	OH	H
oxytetracycline	OTC	H	OH	H	OH
doxycycline	DC	H	H	CH_3	OH
chlortetracycline	CTC	Cl	OH	H	H

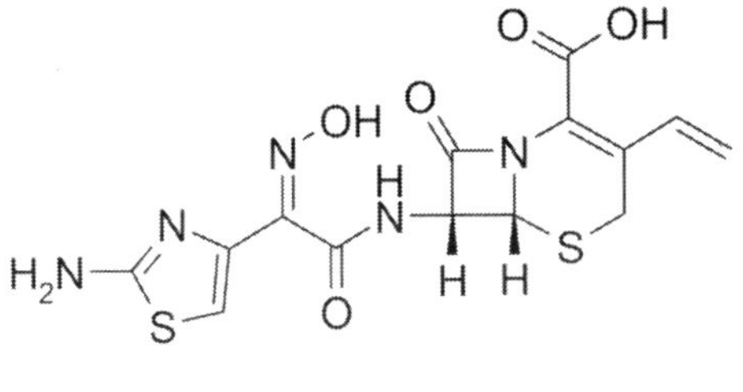

B: cefdinir

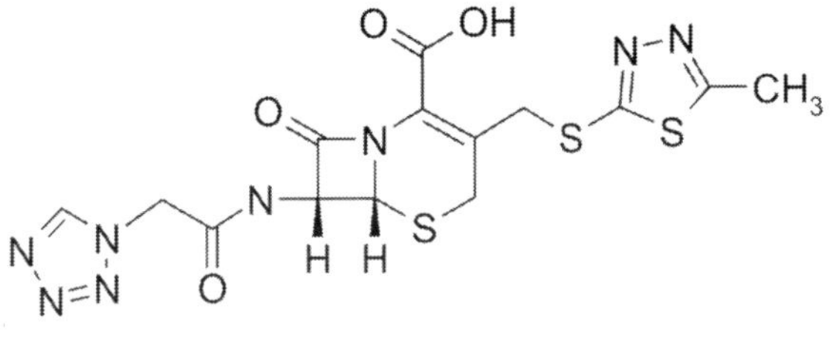

C: cefazoline

Figure 1. Chemical structures of antibiotics investigated.

2.2. *Magnetic seeding and magnetic separation experiment*

In electrochemical magnetic seeding process, the sample solution (100 ml) containing antibiotic and electrolyte (NaCl, 1.0 g) was electrocoagulated in a glass beaker. The electrodes, both anode and cathode were iron rods. The 1.0 A of constant current was applied from DC power supply and the solution was electrocoagulated for 60 seconds. Then 50 mg of Fe_3O_4 (average particle size: 3 μm) was added to 100 ml of electrocoagulated solution and mixed. A neodymium magnet (4,300 gauss) was set under the beaker and left at rest for 5 min. The surface flux density of the neodymium magnet was measured by a gauss meter (F.W. Bell, model 9200). The supernatant was collected to measure the antibiotic concentration. The removal of antibiotic is defined as Cs/C_i, where C_i is antibiotic concentration in the sample solution before electrocoagulation and Cs is in the supernatant.

2.3. *Analytical method*

The concentration of antibiotics was analyzed by capillary electrophoresis (Agilent, G1600) with a diode array detector set at 282 nm. A fused-silica capillary 48.5 cm × 50 μm i.d. was used and the temperature of the capillary was maintained at 25°C. The measurement was obtained under constant voltage of 25 kV. The composition of electrophoresis buffer was 25 m mol/L Tris/Hydrochloride and 15 m mol/L sodium tetraborate [6].

3. Results and Discussion

3.1. *Magnetic separation of antibiotics with electrochemical magnetic seeding with iron anode*

Table 1 shows an experimental result of magnetic separation of antibiotics. More than 80% of tetracycline antibiotics (oxytetracycline, tetracycline and doxycycline) were removed from sample solution. The electrocoagulation includes coagulation, absorption, adsorption and floatation. During the electrochemical process, metal ions served at the sacrificial anode react with water to form metallic oxides and hydroxides. It is known that electrocoagulation with an iron anode is available for magnetic seeding of several substances. Organic matters in aqueous solution are partially adsorbed onto the surface of metallic oxides and hydroxides formed by electrocoagulation. It is also known that high molecular iron chelates are formed with tetracycline antibiotics [7]. Mooney reported that cefdinir forms stable complexes with iron (II) and iron (III) under neural conditions [8]. It is considered that the electrocoagulation with an iron anode forms iron hydroxides and tetracycline-iron or cefdinir-iron complexes. However, the iron hydroxides formed by electrocoagulation may have much lower susceptibility for magnetic separation using a permanent magnet. Tsouris et al. indicated that magnetic susceptibility of electrogenerated particles increases with temperature of solution [5]. To expand the difference in magnetic susceptibility between electrogenerated hydroxides and water, the magnetite powder was added to the solution after the electrocoagulation. The ferromagnetic particles were entrapped immediately into iron colloids formed by electrocoagulation. The antibiotic-iron complexes

Table 1. Magnetic separation of antibiotics with electrochemical magnetic seeding with iron anode

Class	Substance	Initial (mg/L)	After separation (mg/L)	Removal (%)	pH (-)	Iron-chelating
Tetracycline	Oxytetracycline	99.7	10.6	89.3	3.93	+
	Doxycycline	99.2	19.5	80.3	3.92	+
	Chlortetracycline	98.8	21.6	78.1	3.88	+
	Tetracycline	98.5	6.7	93.2	3.89	+
Cephalosporin	Cefdinir	99.4	12.0	88.0	6.42	+
	Cefazolin	99.6	81.0	18.7	7.75	

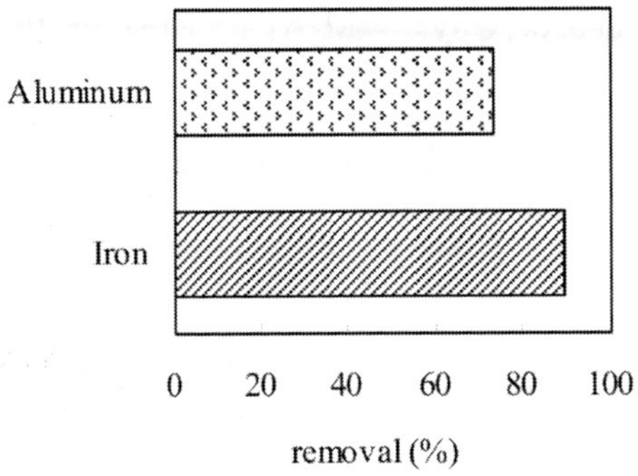

Figure 2. Effect of anode material on electrochemical magnetic seeding for magnetic separation of oxytetracycline (pH: 7.05)

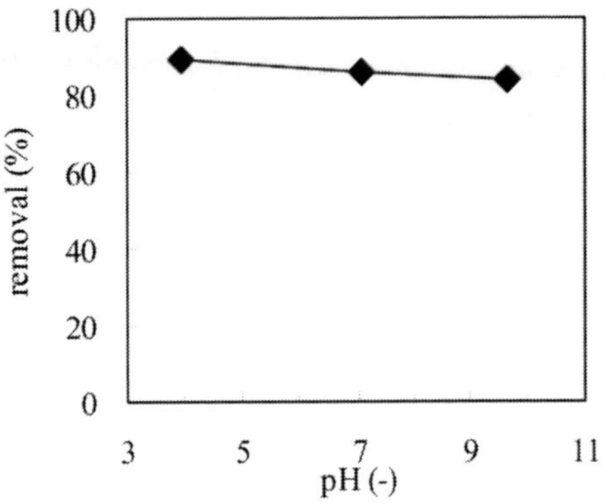

Figure 3. Effect of pH on magnetic seeding for magnetic separation of oxytetracycline.

and magnetite might be adsorbed onto colloidal particles of iron hydroxides. It was indicated that the iron colloids play a role in interfacing antibiotic-iron complexes and ferromagnetic substance. In contrast to tetracycline antibiotics, the removal efficiency of cefazolin classified into cephalosporin antibiotics was relatively low. According to the pharmaceutical interaction, cefazolin can hardly form iron chelates. It was indicated that the iron-antibiotic complex allows higher removal by magnetic separation. The result showed that the magnetic separation with electrocoagulation using iron electrodes is effective method for treatment of tetracycline antibiotics.

3.2. *Effect of condition of electrochemical magnetic seeding*

Fig. 2 shows the effect of anode material on electrochemical magnetic seeding for magnetic separation of oxytetracycline. High removal rates were obtained with both aluminum and iron anodes, since tetracycline compounds are capable of forming complexes with iron and aluminum. Effect of pH on magnetic seeding using iron anode for magnetic separation of oxytetracycline was shown in figure 3. The pH ranged from 3.9 to 9.6 had little effect on removal rate. The result indicated that the electrochemical magnetic seeding method is available under moderately acidic, neutral and moderately alkaline conditions.

4. Conclusion

Electrochemical magnetic seeding of antibiotics using sacrificial anodes was investigated in this work. The result identified that this method is efficient in removing antibiotics, which are capable in forming metal complexes. It should be noted that the formation of metal complex was effective in achieving selective separation for magnetic seeding. For practical application, the separation test for the actual wastewater such as animal manure or wasted milk is required.

Acknowledgement
This work was partly supported by Magnetic Health Science Foundation.

References

[1] Santos S M, Henriques M, Duarte A C, Esteres V I 2007 *Talanta* **71** 731
[2] Seino A, Furusho S and Matsunaga S 2004 *J. Jpn. Society on Water Environ.* **27** 685
[3] Kemper N 2008 *Ecological Indicators* **8** 1
[4] Sarmaha A K, Meyerb M T and Boxall A B A2006 *Chemosphere* **65** 725
[5] Tsouris C, DePaoli D W, Shor J T, Hu M Z C and Ying T Y 2000 *Colloids and Surfaces A: Physicochemical and Engineering Aspects* **177** 223

[6] Fariaa A F, de Souzab M V M, de Almeidaa M V and de Oliveira M A L 2006 *Analytica Chimica Acta* **579** 185
[7] Gu C and Karthikeyan K G 2005 *Environmental Science & Technology* **39** 2660
[8] Mooney M T, Degchi S, Tada T, Fujioka M, Okamoto Y and Yasuda T 1995 *Chemical & pharmaceutical bulletin* **43** 374

AUTHOR INDEX

Akiyama, J 16
Asai, S 16, 29, 40
Awaji, S 110
Barthem, V M T S 45
Beaugnon, E 45, 52
Beneragama, N 188
De Miranda, P E V 45
Farjami, S 57
Fujimura, Y 161
Fujiwara, Y 120, 145
Fukuda, T 57, 61, 70
Fukui, S 175
Garcin, T 45, 52
Givord, D 45
Hagio, T 16
Hagiwara, M 128, 133
Hayakawa, T 80
Hayashi, H 84
Hirota, N 80, 92
Hisayoshi, K 115, 128, 133
Ihara, I 188
Iino, M 161
Ikehata, M 80, 92
Ito, D 185
Iwai, K 9, 16, 29, 40
Iwasaka, M 92, 166
Jang, B K 23
Kakeshita, T 57, 61, 70
Kakuda, Y 171
Kanayama, H 175
Kano, S 115, 128, 133
Kashiwagi, T 128, 133
Kida, T 128, 133
Kimura, F 6
Kimura, T 1, 6
Kiso, K 84
Kohama, T 9
Koyama, K 99, 104
Kuroki, S 84
Lee, J H 70
Matsumoto, Y 84
Matuo, Y 180
Mishima, F 180
Mitsui, Y 99
Miura, O 185
Miyakoshi, J 74
Mogi, I 110, 155
Moribe, S 145
Nakao, R 180
Nikawa, J 84
Nishijima, S 180
Nishimura, K 185
Noto, K 175
Ogawa, J 175
Ogawa, K 34
Oka, T 175
Ooba, T 84
Ooizumi, M 175
Otabe, E S 84
Ozawa, S 34
Rivoirard, S 45, 52
Sakaida, Y 29
Sakakura, R 104
Sakka, Y 23
Sakurai, T 74
Sato, T 175
Suwa, M 137
Taguchi, T 180
Takahashi, K 1, 99, 110
Tanaka, K 175
Tanase, T 16
Tanimoto, Y 120, 145, 171
Toyoda, K 188
Toyoda, N 40
Umetsu, K 188
Usui, M 40
Uyeda, C 115, 128, 133
Wakita, Y 145
Watanabe, K 1, 99, 104, 110, 155
Watarai, H 137
Woo, S K 23
Yamada, S 145
Yamaguchi, M 34, 175
Yamamoto, I 34
Yamato, M 1
Yokoyama, K 175
Yonemura, H 145
Yoshie, S 80

Institute of Physics
Dirac House, Temple Back
Bristol BS1 6BE UK